FLUID DYNAMICS
VIA EXAMPLES AND SOLUTIONS

T0132631

FLUID DYNAMICS
VIA EXAMPLES AND SOLUTIONS

SERGEY NAZARENKO
University of Warwick, UK

CRC Press
Taylor & Francis Group
Boca Raton London New York

CRC Press is an imprint of the
Taylor & Francis Group, an **informa** business

CRC Press
Taylor & Francis Group
6000 Broken Sound Parkway NW, Suite 300
Boca Raton, FL 33487-2742

© 2015 by Taylor & Francis Group, LLC
CRC Press is an imprint of Taylor & Francis Group, an Informa business

No claim to original U.S. Government works

Printed on acid-free paper
Version Date: 20141027

International Standard Book Number-13: 978-1-4398-8882-7 (Paperback)

Visit the Taylor & Francis Web site at
http://www.taylorandfrancis.com

and the CRC Press Web site at
http://www.crcpress.com

Contents

5 Boundary layers   99

6 Two-dimensional flows   115

Preface

As an applied subject, fluid dynamics is best studied via considering specific examples and solving problems dealing with various phenomena and effects in fluids. This is well recognised in most Fluid Dynamics courses, which often have support classes devoted to considering physically motivated exercises. Also, such Fluid Dynamics courses are typically assessed via solution of specific problems more often than via reproducing mathematical proofs or general abstract constructions. However, original Fluid Dynamics problems are rather hard to invent, and the Fluid Dynamics lecturers are often "on their own" having to reinvent successful ideas, tricks and representative examples. The present book addresses these issues by systematically providing such ideas and model examples.

A distinct feature of the present book is that it is problem oriented. Of course, there are many wonderful fluid dynamics textbooks, classical and more recent, which contain exercises and examples, to name just a few: *Hydrodynamics* by H. Lamb [13], *Essentials of Fluid Dynamics* L. Prandtl [21], *An Introduction to Fluid Dynamics* by G.K. Batchelor [4], *Fluid Dynamics* by L.D. Landau and E.M. Lifshitz [14], *Prandtl-Essentials of Fluid Mechanics* by Oertel et al. [17], *Elementary Fluid Dynamics* by D.J. Acheson [1], *Fluid Mechanics* by P.K. Kundu and I.M. Cohen [12], *Physical Fluid Dynamics* by D.J. Tritton [28], *A First Course in Fluid Dynamics* by A.R. Paterson [19], *Elementary Fluid Mechanics* by T. Kambe [9], *Fluid Mechanics: A Short Course for Physicists* by G. Falkovich [6], *Fundamentals of Geophysical Fluid Dynamics* by J.C. McWilliams [16] and *Waves in Fluids* by J. Lighthill [15]. However, most of the existing books make an accent on the theory or expositions. There has been a clear lack of a text which would contain a sizeable set of example problems and detailed model solutions. The present book is intended to fill this gap by presenting a number of fluid dynamics problems organised in chapters dealing with several sub-areas, types of flows and applications. The problems form a "skeleton" of the book structure. Throughout this book, we include supplementary theoretical material when necessary, with an extended list of references for suggested further reading material at the end of each chapter. We also provide a complete set of model solutions.

The book is designed to be used in problem solving support classes and for exam revision in undergraduate and graduate fluid dynamics courses. Also, the book will aid lecturers by offering a pool of possible exam questions for such fluid dynamics courses. It is my hope that the book could be useful also to students and lecturers in related subjects, such as continuum mechanics, turbulence, ocean and atmospheric sciences, etc. More broadly, the provided set of example problems should help an effective hands-on study of fluid dynamics, within or outside of a university course, including an independent study by specialists in other scientific areas who would like to learn basics of fluid dynamics.

Acknowledgements

I am grateful to Jason Laurie for a thorough proofreading of the manuscript. I thank Davide Faranda, Denis Kuzzay, Vadim Nikolaev, and Miguel Onorato for their comments that allowed to improve the presentation. The book was partially written during my sabbatical leave spent at the Institute of Computational Technologies in Novosibirsk, Russia, in 2012–13 and at the SPEC laboratory at the Commissariat à l'Énergie Atomique in Saclay, France, in 2013–14. I am grateful to both of these organisations for their genuine hospitality.

List of Figures

Author Biography

Sergey Nazarenko's research is in the areas of fluid dynamics, turbulence and waves arising in different applications. This includes wave turbulence, magneto-hydrodynamic turbulence, superfluid turbulence, water waves, Rossby waves, vortices and jets in geophysical fluids, drift waves and zonal jets in plasmas, optical vortices and turbulence, turbulence in Bose-Einstein condensates.

Sergey Nazarenko has been working at the University of Warwick since 1996, where presently he holds a position of full professor. Prior to Warwick, Sergey Nazarenko worked as visiting assistant professor at the Department of Mathematics, University of Arizona in 1993–1996, postdoc at the Department of Mechanical and Aerospace Engineering, Rutgers University, New Jersey in 1992–1993, and researcher at the Landau Institute for Theoretical Physics, Moscow in 1991–1992.

Sergey Nazarenko wrote *Wave Turbulence* published by Springer in 2011. He was a co-editor of the books *Non-equilibrium Statistical Mechanics and Turbulence*, CUP 2008, and *Advances in Wave Turbulence*, World Scientific, 2013. Sergey Nazarenko has organised twelve international scientific conferences and workshops in fluid dynamics and turbulence.

Chapter 1

Fluid equations and different regimes of fluid flows

1.1 Background theory

A fluid is a continuous medium whose state is characterised by its velocity field, $\mathbf{u} = \mathbf{u}(\mathbf{x}, t)$, pressure and density fields, $p = p(\mathbf{x}, t)$ and $\rho = \rho(\mathbf{x}, t)$ respectively, and possibly other relevant fields (e.g. temperature). Here, t is time and $\mathbf{x} \in R^d$ is the physical coordinate in the d-dimensional space, $d = 1, 2$ or 3. Respectively, $\mathbf{u} \in R^d$, although in some special flows the dimensions of $\mathbf{x}$ and $\mathbf{u}$ may be different from each other.

Most of the fluid dynamics results have been obtained starting from the *Navier-Stokes equations*. These equations have many variations depending on the forces acting on the fluid, as well as the properties of the fluid itself—compressibility, thermoconductivity, viscosity, density homogeneity/inhomogeneity, chemical composition, etc. We will consider a relatively small but important subset of idealised cases.

1.1.1 Incompressible flows

The *Navier-Stokes equation* for incompressible flow is given by

$$D_t \mathbf{u} = -\frac{1}{\rho}\nabla p + \nu\nabla^2 \mathbf{u} + \mathbf{f}, \tag{1.1}$$

which is a momentum balance equation for fluid particles (a continuous medium version of Newton's second law). We introduced notation for the fluid particle acceleration,

$$D_t \mathbf{u} \equiv \partial_t \mathbf{u} + (\mathbf{u} \cdot \nabla)\mathbf{u}. \tag{1.2}$$

Operator $D_t \equiv \partial_t + (\mathbf{u} \cdot \nabla)$ is a time derivative along the fluid particle trajectory—a Lagrangian time derivative. Possible external forces acting on the fluid (per unit mass) are denoted by term $\mathbf{f}$ in the right-hand side of equation (1.1); these could be gravity, electrostatic force, etc.

The momentum balance equation (1.1) has to be complemented by a mass

1

balance equation, which for an incompressible fluid is

$$\nabla \cdot \mathbf{u} = 0, \tag{1.3}$$

and is usually called the *incompressibility condition.*

Note that fluids can be incompressible and yet the density of the fluid particles may vary in physical space. In this case we need an extra equation describing the conservation of ρ along the fluid particle trajectories,

$$D_t \rho = 0. \tag{1.4}$$

1.1.2   Inviscid flows

When the Reynolds number is large one can ignore viscosity (see problem 1.3.1), and the momentum balance equation (1.1) reduces to

$$D_t \mathbf{u} = -\frac{1}{\rho}\nabla p + \mathbf{f}, \tag{1.5}$$

which is known as the *Euler equation.* It is valid for both compressible and incompressible fluids. However, for compressible fluids, the mass balance equation is now different:

$$\partial_t \rho + \nabla \cdot (\rho \mathbf{u}) = 0. \tag{1.6}$$

(This equation remains the same in presence of viscosity). Also, since there is an extra unknown field, $\rho(\mathbf{x}, t)$, we need an extra evolution equation for the model to be complete. Generally, such an equation is provided by the energy balance relation. In particular, assuming that different fluid particles are thermally insulated from each other, one can write the additional equation in the form of conservation of *entropy* $S \equiv S(\mathbf{x}, t)$ along the fluid paths,

$$D_t S = 0. \tag{1.7}$$

For the polytropic gas model

$$S = C_v \ln \frac{p}{\rho^\gamma}, \tag{1.8}$$

where constants C_v and γ are called the specific heat constant and the adiabatic index respectively. Obviously, C_v drops out of the equation (1.8) and therefore it is irrelevant in this case. For monatomic ideal gas (e.g. helium, neon, argon) $\gamma = 5/3$; for diatomic gas $\gamma = 7/5$ (e.g. oxygen, nitrogen).

In the simplest case of *isentropic gas*, $S =$const, i.e.

$$p \propto \rho^\gamma. \tag{1.9}$$

In incompressible fluids the equation (1.7) implies conservation of temperature T along the fluid paths,

$$D_t T = 0. \tag{1.10}$$

Moreover, in this case the flow velocity field is not affected by the temperature distribution, i.e. the temperature is *passively advected* by the flow.

In presence of thermal conductivity, the fluid particles are no longer insulated from each other and equation (1.10) (for an incompressible flow) should be replaced by

$$D_t T = \kappa \nabla^2 T, \qquad (1.11)$$

where κ is the thermal conductivity coefficient. In presence of viscosity one must also add a heat production term due to the internal viscous friction, but this contribution can be neglected if the temperature is high and the velocity gradients are small.

1.1.3 Rotating flows

In rotating fluids, it is convenient to write the Navier-Stokes equations in a rotating (rather than the inertial/laboratory) frame of reference. In the frame rotating with an angular velocity $\boldsymbol{\Omega}$ the Navier-Stokes equations become

$$D_t \mathbf{u} = -\frac{1}{\rho} \nabla p_R - 2\boldsymbol{\Omega} \times \mathbf{u} + \nu \nabla^2 \mathbf{u} + \mathbf{f}, \qquad (1.12)$$

where $p_R = p - \Omega^2 r^2/2$ is the so-called reduced pressure and r is the distance from $\mathbf{x}$ to the rotation axis. Term $-2\boldsymbol{\Omega} \times \mathbf{u}$ is called Coriolis acceleration.

1.2 Further reading

Discussions of various dynamical regimes in fluids can be found in most Fluid Dynamics textbooks, e.g. in the books *Elementary Fluid Dynamics* by D.J. Acheson [1], *Fluid Dynamics* by L.D. Landau and E.M. Lifshitz [14] and *Elementary Fluid Mechanics* by T. Kambe [9]. More advanced discussions of regimes with rotation and stratification relevant to the geophysical flows can be found in the book *Fundamentals of Geophysical Fluid Dynamics* by J.C. McWilliams [16].

1.3 Problems

The determination of the regime a fluid is in, and the choice of which simplifications of the full set of equations can be made, are based on a few non-dimensional numbers characterising a particular flow. The problems below consider a few examples of such numbers: the Reynolds number, the Mach number, the Rossby number, the Richardson number, the Prandtl number and the Stokes number.

1.3.1   Reynolds number

Consider a flow with a typical scale of variation of its velocity field L and typical velocity magnitude U. For what values of L, U and the viscosity coefficient ν can one neglect the influence of viscosity? For what values of these parameters can one ignore the fluid particle acceleration? Express both answers in terms of the conditions on the Reynolds number

$$Re = \frac{UL}{\nu}. \tag{1.13}$$

1.3.2   Mach number

Consider a compressible isentropic flow with a typical velocity magnitude U. For what values of U can the compressibility of the fluid be neglected? Express your answer in terms of the Mach number,

$$M = \frac{U}{c_s}, \tag{1.14}$$

where

$$c_s = \sqrt{\frac{\partial p}{\partial \rho}} \tag{1.15}$$

is the speed of sound.

1.3.3   Rossby number

Consider an incompressible flow with a typical velocity U and typical length scale L in a uniformly rotating (with angular velocity $\boldsymbol{\Omega}$) frame of reference.

1. For what values of U, L and $\Omega = |\boldsymbol{\Omega}|$ can one neglect the nonlinear term $(\mathbf{u} \cdot \nabla)\mathbf{u}$? Express your answer in terms of the Rossby number,

$$Ro = \frac{U}{\Omega L}. \tag{1.16}$$

2. *Taylor-Proudman theorem.* Consider a steady incompressible inviscid flow in a uniformly rotating frame of reference with $Ro \ll 1$. Show that in the leading order in small Ro, the velocity and the pressure fields are independent of the coordinate projection to the rotation axis.

1.3.4   Richardson number

Consider a fluid flow in a gravity field whose density is stratified with a mean profile $\rho(z)$ and a typical size in the vertical direction h. If the kinetic energy of fluid particles in such a flow exceeds the work done by the buoyancy force when it is moved a distance $\sim h$ in the vertical direction, then the buoyancy effect may be neglected and the fluid will quickly homogenise. In the opposite case, when the stratification is stable, $\partial_z \rho(z) < 0$, and when the negative work needed to be done by the buoyancy force to move a fluid element vertically a distance $\sim h$ is greater than the kinetic energy, the fluid element will not be able to move over distance $\sim h$ and the stratification will persist with the same mean profile $\rho(z)$. When the stratification is unstable, $\partial_z \rho(z) > 0$, then the flow will be buoyancy driven, i.e. it will gain the kinetic energy due to the positive work done by the buoyancy force.

The Richardson number Ri is a dimensionless number, the value of which determines whether the buoyancy force is an important factor governing the dynamics of a flow. Namely, this number is a measure for the typical ratio of the absolute value of work done by the buoyancy force on a fluid particle to the kinetic energy of this particle.

1. Find the buoyancy (Archimedes) force on a fluid element with infinitesimal volume V and density ρ_1 which is surrounded by fluid of density ρ_2.

2. Find the work done by the buoyancy force on a fluid element when it is moved vertically a distance h.

3. Define the Richardson number Ri as the typical ratio of the absolute value of work done by the buoyancy force on a fluid particle to the kinetic energy of this particle and express it in terms of the typical vertical length h, typical gradient of the density gradient $\rho' = \partial_z \rho(z)$, typical velocity U and the gravity acceleration g.

4. Describe the character of the fluid motion when the initial Richardson number is much greater than unity both in the case of the stable and the unstable stratifications. Will the Richardson number remain much greater than unity? Estimate the characteristic time constants in the respective cases.

1.3.5   Prandtl number

The Prandtl number Pr is a dimensionless number defined as

$$Pr = \frac{\nu}{\kappa},$$

where ν is the kinematic viscosity coefficient, and κ is the thermal conductivity coefficient.

Typical values are: $Pr \ll 1$ for liquid metals (thermal diffusivity dominates), $Pr \gg 1$ for motor oil (momentum diffusivity dominates), $Pr \sim 1$ for air and $Pr \sim 10$ for water.

1. Consider a flow of viscous heat-conducting fluid over a semi-infinite flat plate; see figure 5.2. The plate is hotter than the ambient flow. Two layers will form at the plate: a thermal boundary layer defined by the distance from the plate at which the temperature relaxes to the ambient temperature and a velocity boundary layer defined by the distance at which the flow velocity relaxes to the velocity value at infinity. Which of the two layers will be thicker if the fluid is mercury? Air? Water? Motor oil? Assume that the heat production due to viscosity is significantly less than the heat supplied by the plate.

2. Apart from the temperature, the Prandtl number is also used to quantify passive advection-diffusion of material substances, e.g. pollutant particles. In this case κ is the pollutant's diffusion coefficient. In turbulent flows, due to random mixing, the pollutant concentration acquires small-scale structure.

   How does the smallest scale in the pollutant concentration field compare with the size of the smallest vortices in turbulence for $Pr \ll 1$? For $Pr \sim 1$? For $Pr \gg 1$? Explain your answers.

3. Estimate the ratio of the smallest scale in the pollutant concentration field, ℓ_p, to the size of the smallest vortices in turbulence, ℓ_u, as a function of Pr for $Pr \gg 1$ (the so-called Batchelor's regime of the passive scalar turbulence). **Hint**: the leading contribution to the pollutant's advection in this regime comes from the velocity gradients produced by the smallest turbulent vortices.

1.3.6 Stokes number

The Stokes number St is a dimensionless number quantifying the behaviour of particles suspended in a fluid flow. It is defined as

$$St = \sigma\tau,$$

where τ is the particle's relaxation time to the velocity of the fluid flow (due to a drag), and σ is a typical velocity gradient in the flow (shear or strain). Note that $1/\sigma$ is a typical variation time of the flow velocity field in the frame co-moving with a fluid particle. Particles with a low Stokes number closely follow the fluid elements. This is used in the particle image velocimetry (PIV) experimental technique to measure the flow velocity field. For a large Stokes number, the particle's inertia dominates so that the particle will continue moving with its initial velocity along a nearly straight line (think e.g. of a large rain droplet).

1. When the Reynolds number based on the particle size d is low, e.g. when the particle is small, the flow around it will be laminar and with a single length scale $\sim d$. Moreover, when the particle is neutrally buoyant (i.e. with the same density as the flow) the relaxation time cannot depend on the density. Use a dimensional argument to find an estimate for the relaxation time τ in terms of d and the kinematic viscosity coefficient ν. Write the Stokes number in terms of the same quantities and σ.

2. Suppose that the Stokes number is low, $St \ll 1$. Consider a moving particle in a 2D vortex with circular streamlines and a particle near a stagnation point where the streamlines are nearly hyperbolic. Describe qualitatively the particle motion in each of the two cases. Use these examples and generalise your conclusions to the particle in a flow which involves many vortices and stagnation points in between them.

3. Suppose now that the Stokes number is high, $St \gg 1$. Consider a particle that moves with velocity U through turbulent air with characteristic strain s and typical size of vortices L. Find the mean distance to which the particle deviates in the transverse to its main motion direction at time t. **Hint**: consider a simplified model in which the particle makes one random step in the transverse plain each time it travels through one fluid vortex.

1.4   Solutions

1.4.1   Model solution to question 1.3.1

Typical value of the fluid particle acceleration is

$$|(\mathbf{u} \cdot \nabla)\mathbf{u}| \sim U^2/L,$$

whereas the typical value of the viscous term is

$$|\nu\nabla^2\mathbf{u}| \sim \nu U/L^2.$$

The ratio of these two typical values makes the Reynolds number $Re = \frac{UL}{\nu}$. Thus, one can neglect viscosity if $Re \gg 1$, whereas the fluid particle acceleration term can be ignored if $Re \ll 1$.

1.4.2   Model solution to question 1.3.2

Balancing the inertia term $(\mathbf{u} \cdot \nabla)\mathbf{u}$ with the pressure force term in the Euler equation (1.5), we obtain an estimate for relative changes in density, $\delta\rho$:

$$(\mathbf{u} \cdot \nabla)\mathbf{u} \sim \frac{\nabla p}{\rho} \quad \rightarrow \quad U^2/L \sim \frac{\partial p}{\partial \rho}\frac{\delta\rho}{\rho L} \quad \rightarrow \quad \frac{\delta\rho}{\rho} \sim \frac{U^2}{c_s^2} = M^2.$$

Therefore, the fluid compressibility is negligible ($\delta\rho \ll \rho$) when the Mach number is small, $M \ll 1$.

1.4.3 Model solution to question 1.3.3

1. In the equation (1.12), let us estimate the nonlinear inertial term as $|(\mathbf{u}\cdot\nabla)\mathbf{u}| \sim U^2/L$ and the Coriolis term as $|-2\,\mathbf{\Omega}\times\mathbf{u}| \sim \Omega U$. Thus, the nonlinear inertial term is much less than the Coriolis term when

$$Ro = \frac{U}{\Omega L} \ll 1. \tag{1.17}$$

2. The equation (1.12) with $\nu = \mathbf{f} = 0$ becomes:

$$2\rho\mathbf{\Omega}\times\mathbf{u} = -\nabla p_R \tag{1.18}$$

(here one could also add a potential force which would only redefine p_R). From the z-component (parallel to $\mathbf{\Omega}$) to of this equation we get

$$\partial_z p_R = 0.$$

Applying the curl operator to equation (1.18) and taking into account that $\nabla\cdot\mathbf{u} = 0$, we arrive at

$$(\mathbf{\Omega}\cdot\nabla)\mathbf{u} = 0, \quad \text{or} \quad \partial_z\mathbf{u} = 0.$$

Both the pressure and the velocity fields are independent of the coordinate projection to the rotation axis. Thus we have proven the Taylor-Proudman theorem.

1.4.4 Model solution to question 1.3.4

1. According to the Archimedes law, the buoyancy force F on a fluid element with infinitesimal volume V and density ρ_1 which is surrounded by fluid of density ρ_2 is

$$F = (\rho_2 - \rho_1)gV.$$

2. The work done by the buoyancy force F on a fluid element when it is moved vertically a distance h is

$$A = \int_0^h F(z)\,dz = gV\int_0^h [\rho(z) - \rho(0)]\,dz \sim gV\rho'h^2.$$

3. The Richardson number defined as the typical ratio of the absolute value of work done by the buoyancy force on a fluid particle to the kinetic energy of this particle is:

$$Ri = \frac{gV\rho'h^2}{\rho V u^2} = \frac{g\rho'h^2}{\rho u^2}.$$

4. In the case of the stable stratification, the particles will oscillate around their equilibrium positions. The Richardson number in this case will remain much greater than unity and the characteristic time τ will simply be the wave period, which can be found from the equation of motion of the fluid particle:

$$\rho V \ddot{z}(t) = F = gV\rho' z(t),$$

i.e.

$$\tau = \frac{2\pi}{\omega} = \frac{2\pi}{\sqrt{g|\rho'|/\rho}}.$$

In the unstable stratification case (heavy fluid on top of a light one), the instability will result in the light particles rising to the top and accelerating. The available potential energy will get converted into the kinetic energy. Thus, the Richardson number in this case will be lower, a value around unity. The characteristic time constant will be the inverse growth rate of the instability

$$\tau = \frac{1}{\gamma} = \frac{1}{\sqrt{g|\rho'|/\rho}}.$$

1.4.5 Model solution to question 1.3.5

1. For mercury, $Pr \ll 1$, so the heat diffusion is faster than the momentum diffusion. Hence the thermal boundary layer will be thicker than the velocity boundary layer. For air $Pr \sim 1$, i.e. the heat diffusion occurs at a similar rate as the momentum diffusion. Hence the thermal boundary layer will be of approximately the same thickness as the velocity boundary layer. For water and motor oil $Pr \gg 1$, so the heat diffusion is slower than the momentum diffusion. Hence the thermal boundary layer will be thinner than the velocity boundary layer (a lot thinner in the case of the motor oil).

2. The smallest vortex scale in turbulence, ℓ_u, (Kolmogorov scale) will be determined by the balance of the advection time-scale and the viscous time-scale. The smallest scale in the pollutant concentration field, ℓ_p, will be determined by the balance of the advection time-scale and the diffusion time-scale. Thus, the smallest scale in the pollutant concentration field ℓ_p will be much greater than the smallest vortex size ℓ_u for $Pr \ll 1$, of the same order for $Pr \sim 1$, and much less for $Pr \gg 1$.

3. In Batchelor's regime of the passive scalar turbulence, when $Pr \gg 1$, the typical advection time-scale is the same for both the smallest turbulent vortex and for the scalar field at the scales smaller than ℓ_u. This typical advection time-scale is $\tau_u = 1/u'$, where u' is the typical velocity gradient produced by the smallest turbulent vortices. The smallest scale in the pollutant concentration field is determined by the balance of τ_u

and the scalar diffusion time ℓ_p^2/κ. The size of the smallest vortices in turbulence is determined by the balance of τ_u and the viscous diffusion time ℓ_u^2/ν. Indeed, the smaller pollutant structures would quickly diffuse and become bigger. Therefore $\ell_p^2/\kappa \sim \ell_u^2/\nu$, or finally,

$$\frac{\ell_u}{\ell_p} = \sqrt{Pr}.$$

1.4.6   Model solution to question 1.3.6

1. When the Reynolds number based on the particle size d is low, e.g. when the particle is small, the flow around it will be laminar and with a single length scale $\sim d$. Moreover, when the particle is neutrally buoyant (i.e. with the same density as the flow) the relaxation time cannot depend on the density. Using the dimensional argument we find

$$\tau = \frac{Cd^2}{\nu},$$

where C is a dimensionless constant (which is, based on a more rigorous analysis, equal to $1/18$).

Respectively, the Stokes number is

$$St = \frac{Cd^2\sigma}{\nu}.$$

2. Due to a finite relaxation time, there is some delay in adjusting the particle's velocity to the one of the flow, in particular in adjusting its direction. Because of this, the particle's velocity is not tangential to the flow—it has a finite (small if $St \ll 1$) negative normal component with respect to the streamline. For a particle moving in a 2D vortex with circular streamlines, it means spiralling away of the vortex centre, and for a particle near a stagnation point it means moving closer to the separatrices (streamlines passing through the stagnation point). Generalising, the particle in a flow which involves many vortices and stagnation points tends to be pushed away from the vortices into the in-between space adjacent to the separatrices.

3. When the Stokes number is high, $St \gg 1$, the particle moves mostly in one direction with small deviations in the transverse plane because of the fluid vortices. Passing one such vortex would take time $\tau_p = L/U$ and it would result in a transverse deviation $\delta \sim L\sigma\tau_p/St$. Assuming that each of the deviations caused by different vortices are random and statistically independent, for a mean transverse distance $\Delta(t)$ we have a random-walk result:

$$\Delta(t)^2 \sim \frac{\delta^2}{\tau_p}t \sim \frac{\sigma^2 L^3}{U\,St^2}t.$$

Chapter 2

Conservation laws in incompressible fluid flows

2.1 Background theory

Many properties of fluid flows can be understood based on very general conservation laws. Some of these laws are universal in physics, such as the energy and the momentum conservation laws. Each of them represents a global conservation property, i.e. the *total* amount of energy and momentum remain unchanged in the system. The other type of conservation laws in fluids are *local or Lagrangian*: they refer to conservation of a field along fluid trajectories (e.g. the vorticity in 2D) or conservation over a selected set of moving fluid particles (e.g. the velocity circulation over contours made out of moving fluid particles).

2.1.1 Velocity-vorticity form of the Navier-Stokes equation

Let us consider the Navier-Stokes equation for incompressible flow (1.1) under gravity forcing,

$$(\partial_t + (\mathbf{u} \cdot \nabla))\mathbf{u} = -\frac{1}{\rho}\nabla p - g\,\hat{\mathbf{z}} + \nu\nabla^2\mathbf{u}. \tag{2.1}$$

Let us define the vorticity $\boldsymbol{\omega}$ as

$$\boldsymbol{\omega} = \nabla \times \mathbf{u} \tag{2.2}$$

and use vector identity

$$(\mathbf{u} \cdot \nabla)\mathbf{u} = \boldsymbol{\omega} \times \mathbf{u} + \nabla\left(\frac{u^2}{2}\right),$$

which is valid for incompressible flows, i.e. $\nabla \cdot \mathbf{u} = 0$.

Then equation (2.1) can be rewritten as

$$\partial_t\mathbf{u} + \boldsymbol{\omega} \times \mathbf{u} = -\nabla B + \nu\nabla^2\mathbf{u}, \tag{2.3}$$

where we have introduced the *Bernoulli potential*:

$$B = \frac{p}{\rho} + \frac{u^2}{2} + gz. \tag{2.4}$$

2.1.2   Bernoulli theorems

Irrotational flows are defined as the flows with zero vorticity field, $\boldsymbol{\omega} = 0$. For the irrotational flows, the velocity field can be represented as a gradient of a velocity potential, $\mathbf{u} = \nabla\phi$. For irrotational flow, the second term on the left-hand side of equation (2.3) is zero, and the viscous term is zero too by the incompressibility condition, $\nabla^2\phi = \nabla \cdot \mathbf{u} = 0$. (Another way to see this is to realise that for incompressible fields, $\nabla^2\mathbf{u} = -\nabla \times \boldsymbol{\omega}$). Thus, this equation can be integrated once over a path in $\mathbf{x}$. This gives *Bernoulli theorem for time-dependent irrotational flow*:

$$\partial_t\phi + B = C, \tag{2.5}$$

where C is a constant.

For a *stationary irrotational flow*, when all the fields are time independent at each fixed point $\mathbf{x}$, Bernoulli theorem becomes,

$$B \equiv \frac{p}{\rho} + \frac{u^2}{2} + gz = C. \tag{2.6}$$

Note that C is the same constant throughout the irrotational flow volume.

There is yet another form of the Bernoulli theorem: the one for a steady inviscid flow with non-zero vorticity. Considering the dot product of the inviscid ($\nu = 0$) stationary ($\partial_t\mathbf{u} = 0$) version of equation (2.3) with $\mathbf{u}$ we have,

$$(\mathbf{u} \cdot B) = 0.$$

But $(\mathbf{u} \cdot B)$ is just a steady-state version of the time derivative of B along a fluid path. Thus:

In steady ideal flow, the Bernoulli potential B is constant along the flow streamlines.

Note that for flows with non-zero vorticity field this constant is generally different for different streamlines, whereas for irrotational flows the constant is the same throughout the flow, i.e. it is the same for all streamlines as expressed in equation (2.6).

2.1.3   The vorticity form of the flow equation

Taking the curl of the equation (2.3), assuming $\rho = \text{const}$ or $p \equiv p(\rho)$ (barotropic flow), we have

$$\partial_t\boldsymbol{\omega} + \nabla \times (\boldsymbol{\omega} \times \mathbf{u}) = +\nu\nabla^2\boldsymbol{\omega}. \tag{2.7}$$

Note that this equation is also valid for compressible isentropic fluids, i.e. when the pressure is also a function of the density only. In question 2.3.1 we will consider a generalisation of this equation to baroclinic (non-barotropic) flows and will derive Ertel's theorem.

Using the vector identity

$$\nabla \times (\boldsymbol{\omega} \times \mathbf{u}) = (\mathbf{u} \cdot \nabla))\boldsymbol{\omega} - (\boldsymbol{\omega} \cdot \nabla)\mathbf{u},$$

(which is valid for incompressible flows, $\nabla \cdot \mathbf{u} = 0$) we have the following vorticity form of the flow equation,

$$D_t\boldsymbol{\omega} \equiv (\partial_t + (\mathbf{u} \cdot \nabla))\boldsymbol{\omega} = (\boldsymbol{\omega} \cdot \nabla)\mathbf{u} + \nu\nabla^2\boldsymbol{\omega}. \tag{2.8}$$

The left-hand side of this equation describes the time derivative of the vorticity along a fluid path; hence the notation $D_t\boldsymbol{\omega}$. The first term on the right-hand side is the so-called vortex stretching (it is absent in 2D), whereas the last term on the right-hand side is the term associated with the diffusion of vorticity.

2.1.4 Energy balance and energy conservation

Let us suppose that the fluid is contained in a finite volume whose walls are impenetrable and stationary, so that the normal component of velocity $\mathbf{u}$ is zero at the boundary. For finite viscosity ν, the parallel component of $\mathbf{u}$ is also zero at the boundary—this is the *no-slip boundary condition*: $\mathbf{u} = 0|_{\partial V}$, where ∂V denotes the bounding surface for the volume V occupied by the fluid. However, we will also include the case with zero ν, in which case one does not have a condition on the parallel velocity at the boundary—this is the *free-slip boundary condition*: $(\mathbf{u} \cdot \mathbf{n}) = 0|_{\partial V}$, where $\mathbf{n}$ is a unit vector normal to the boundary.

Dot multiply equation (2.3) by $\mathbf{u}$ (the second term on the left-hand side will vanish) and integrate over the containing volume:

$$\partial_t \int_V \frac{1}{2}u^2 \, d\mathbf{x} = -\int_V \mathbf{u} \cdot \nabla B \, d\mathbf{x} + \nu \int_V \mathbf{u} \cdot \nabla^2\mathbf{u} \, d\mathbf{x}. \tag{2.9}$$

Using the incompressibility condition $\nabla \cdot \mathbf{u} = 0$ in the first term on the right-hand side of this equation, and integrating by parts the last term on the right, we get

$$\partial_t \int_V \frac{1}{2}u^2 \, d\mathbf{x} + \int_V \nabla \cdot (\mathbf{u}B) \, d\mathbf{x} = -\nu \sum_{i,j=1}^{3} \int_V (\nabla_i u_j)^2 \, d\mathbf{x}. \tag{2.10}$$

While integrating by parts in the viscous term, we have dropped the boundary terms because they are zero by virtue of the no-slip boundary conditions (in the case when $\nu = 0$, this boundary condition is incorrect, but then there is no viscous term to start with).

Now, using Gauss's theorem one can write

$$\int_V \nabla \cdot (\mathbf{u}B) \, d\mathbf{x} = \int_{\partial V} B \, (\mathbf{u} \cdot d\mathbf{s}), \tag{2.11}$$

which gives zero, because the normal to the boundary velocity is zero in both the no-slip and the free-slip boundary conditions, $\mathbf{u} \cdot d\mathbf{s} = 0$. (Here, $d\mathbf{s}$ is the oriented infinitesimal surface element whose absolute value is the area of this element and the direction is along the normal to the surface).

Defining the total kinetic energy as

$$E = \frac{\rho}{2} \int_V u^2 \, d\mathbf{x}, \tag{2.12}$$

we arrive at the energy balance equation

$$\dot{E} = -\nu\rho \sum_{i,j=1}^{3} (\nabla_i u_j)^2 \, d\mathbf{x}. \tag{2.13}$$

In the inviscid case, this becomes the energy conservation law

$$\dot{E} = 0. \tag{2.14}$$

It is interesting that in the incompressible inviscid fluid, it is the kinetic energy which is conserved, and not only the total energy, which includes both the kinetic and the potential energies. This is because in incompressible fluids the total potential energy is also conserved separately: a fluid parcel moving upwards and gaining potential energy will always be accompanied by some other parcel of equal volume moving down and loosing the same amount of potential energy.

2.1.5 Momentum balance and momentum conservation

First of all, recall that equation (2.1) is already the momentum balance equation: it was obtained by applying Newton's second law to a moving infinitesimal fluid parcel. For an arbitrary *fixed* volume V occupied by the fluid, we can get a balance equation for the total momentum,

$$\mathbf{M} = \int_V \rho \mathbf{u} \, d\mathbf{x},$$

by integrating equation (2.1) over this volume.

$$\partial_t M_i = -\rho \int_V \mathbf{u} \cdot \nabla u_i \, d\mathbf{x} - \int_V \nabla_i (p + g\rho z) \, d\mathbf{x} + \nu\rho \int_V \nabla^2 u_i \, d\mathbf{x}. \tag{2.15}$$

Using the incompressibility condition and Gauss's theorem, we have for the i-component

$$\partial_t M_i = -\rho \int_{\partial V} \mathbf{u} u_i \, d\mathbf{s} - \int_{\partial V} (p + g\rho z) \, ds_i + \nu\rho \int_{\partial V} (\nabla u_i) \, d\mathbf{s}, \tag{2.16}$$

where ∂V denotes the bounding surface for the volume V and ds_i is the i-component of the oriented surface element.

For steady inviscid flows, equation (2.16) reduces to

$$\rho \int_{\partial V} \mathbf{u} u_i \, ds = - \int_{\partial V} (p + g\rho z) \, ds_i, \tag{2.17}$$

a relation which will be used in several problems below.

If V is the entire volume of the fluid, then obviously $\mathbf{M} = 0$ and the first term on the right-hand side of equation 2.16 is zero by the (no-slip or free-slip) boundary conditions. Then we find that the net effect of the pressure force, the gravity force, and the viscous stress at the boundary must be zero,

$$- \int_{\partial V} (p + g\rho z) \, ds_i + \nu \rho \int_{\partial V} (\nabla u_i) \, ds = 0. \tag{2.18}$$

2.1.6 Circulation: Kelvin's theorem

Let us define the velocity circulation Γ over a closed contour C via the following contour integral,

$$\Gamma = \oint_C \mathbf{u} \cdot d\boldsymbol{\ell} = 0. \tag{2.19}$$

By Stokes theorem, one can rewrite expression (2.19) as a surface integral of the vorticity over the surface S spanned by the contour C,

$$\Gamma = \int_S \boldsymbol{\omega} \cdot d\mathbf{s} = 0. \tag{2.20}$$

Consider a material contour C, the points of which move together with the fluid particles. Now without proof (which can be found in any fluid dynamics textbook) we present **Kelvin's circulation theorem:** In an ideal fluid flow (i.e. inviscid and incompressible), circulation Γ over any material contour C is conserved, i.e. $\dot{\Gamma} = 0$.

2.1.7 Vorticity invariants in 2D flows

Let us consider a 2D flow: $\mathbf{u} = (u(x, y, t), v(x, y, t), 0)$, $\boldsymbol{\omega} = (0, 0, \omega(x, y, t))$. Then the vortex stretching term in the vorticity equation (2.8) is zero. The z-component of this equation becomes:

$$(\partial_t + (\mathbf{u} \cdot \nabla))\omega = \nu \nabla^2 \omega. \tag{2.21}$$

For inviscid flow, this equation becomes

$$(\partial_t + (\mathbf{u} \cdot \nabla))\omega = 0, \tag{2.22}$$

which describes the conservation of vorticity along fluid trajectories.

Such a conservation for each fluid particle results in an infinite number of conservation laws. For example, for any function $f(\omega)$ the following quantity will be conserved,

$$\int f(\omega)\,d\mathbf{x},\qquad(2.23)$$

(provided that this integral converges).

In particular, we can choose $f(\omega) = |\omega|^n$, $(n = 1, 2, 3, ...)$, in which case we get so-called *enstrophy series of invariants*:

$$I_n = \int |\omega|^n\,d\mathbf{x}.\qquad(2.24)$$

The most important of these, in particular for the theory of 2D turbulence, is the invariant for $n = 2$, which is called the *enstrophy*:

$$Z \equiv I_2 = \int \omega^2\,d\mathbf{x}.\qquad(2.25)$$

Later in the chapter we will mostly consider 3D flows, but we will use the results for the conservation laws in 2D later in the chapters devoted to 2D flows, point vortices and turbulence (in the part considering 2D turbulence).

2.2   Further reading

Discussions and derivations of the conservation laws in fluids, including the Kelvin circulation theorem (the derivation of which we have omitted) can be found, e.g., in the following textbooks: *Hydrodynamics* by H. Lamb [13], *Vortex Dynamics* by P.G. Saffman [23], *Elementary Fluid Dynamics* by D.J. Acheson [1], *Fluid Dynamics* by L.D. Landau and E.M. Lifshitz [14], and *Elementary Fluid Mechanics* by T. Kambe [9]. Also highly recommended complementary reading for this part is the book *Prandtl's Essentials of Fluid Mechanics* by Oertel et al. [17].

2.3   Problems

2.3.1   Conservation of potential vorticity

Given information:

- The following relation is valid for any scalar field s and solenoidal (divergence-free) vector fields $\mathbf{a}$ and $\mathbf{b}$,

$$\nabla s \cdot (\nabla \times (\mathbf{a} \times \mathbf{b})) = \mathbf{b} \cdot \nabla(\mathbf{a} \cdot \nabla s) - \mathbf{a} \cdot \nabla(\mathbf{b} \cdot \nabla s). \qquad (2.26)$$

- The Euler equation describing dynamics of inviscid fluids can be written in a mixed velocity-vorticity form as follows,

$$\partial_t \mathbf{u} + \boldsymbol{\omega} \times \mathbf{u} = -\frac{1}{\rho}\nabla p - \nabla \frac{u^2}{2}. \qquad (2.27)$$

Consider an incompressible inviscid fluid. Consider also a "passive" scalar field $s(\mathbf{x}, t)$, that is, such a field that does not change along the trajectories of the fluid particles.

1. Write down the incompressibility condition in terms of the velocity field. Why does the vorticity field satisfy the same equation?

2. Express mathematically the fact of conservation of a field $s(\mathbf{x}, t)$ along the trajectories of fluid particles.

3. In barotropic flows, the pressure depends only on the density via a given function $p = p(\rho)$. Flows that are not barotropic are called baroclinic. Starting from the Euler equation (2.27), derive the vorticity evolution equation for a baroclinic flow (c.f. the vorticity equation (2.7) for the barotropic flow).

4. Suppose that the passive scalar $s(\mathbf{x}, t)$ depends on $\mathbf{x}$ and t only via p and ρ, $s = s(p, \rho)$ (for example, s could be temperature or entropy). Show that for both barotropic and baroclinic fluids

$$\nabla s \cdot (\nabla\rho \times \nabla p) = 0.$$

5. Prove that the field $\lambda = \nabla s \cdot \boldsymbol{\omega}$ is conserved along the trajectories of the fluid particles (this theorem was first proven by Ertel in 1942 [24]).

2.3.2   Tap water

A jet of water of diameter d_0 emerges with velocity u_0 from a tap which is located at height h above a solid horizontal surface, as shown in figure 2.1.

For simplicity, one can assume that water is inviscid and its velocity profile at the tap is flat, i.e. velocity of the emerging jet is independent of the distance from its centre.

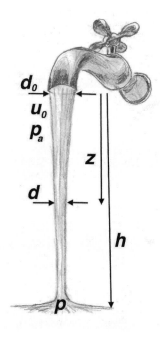

FIGURE 2.1: Tap water.

1. Consider the part of the jet which is of sufficient distance away from the point of impact with the solid surface, where we can assume that velocity remains independent of the distance from the jet's centre. Find the jet velocity u and its diameter d as function of the distance z from the tap.

2. Now consider the impact part of the flow in the vicinity of the solid plane. Find the pressure p at the impact point with the solid surface at the centre of the jet; see figure 2.1.

2.3.3   Discharge into a drainage pipe

A barrel filled with water to level h is connected at its bottom to a horizontal drainage pipe of diameter d and length l, as shown in figure 2.2. A valve is located at the farthest point from the barrel end of the pipe with maximal opening equal to the full cross-section of the pipe.

At time $t = 0$, the valve is changed from the fully locked to the fully open position. For simplicity, we will ignore viscosity and assume that the pipe flow velocity is uniform (independent of the distance from the centre of the pipe).

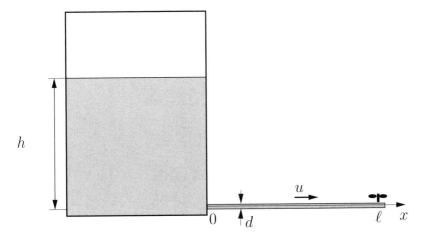

FIGURE 2.2: Drain pipe.

1. Will velocity u depend on the distance x along the pipe?

2. Find the steady-state velocity in the pipe which forms at large time, $u_\infty = \lim_{t \to \infty} u$.

3. Find an expression for the change of velocity u in the pipe as a function of time.

4. Find the pressure p in the pipe as a function of t and x.

2.3.4 Oscillations in a U-tube

Consider the motion of an ideal fluid in a U-shaped tube with constant cross-section; see figure 2.3. At time $t = 0$, the fluid is at rest and the heights of the fluid levels in the two parts of the U-tube are $h_1(0)$ and $h_2(0)$. At $t > 0$, the fluid starts moving in an oscillatory way flowing repeatedly from one part of the U-tube to the other and back. For simplicity, we will assume that the tube flow velocity is independent of the distance from the centre of the U-tube. We will also assume that the round part of the U-tube is small compared to the two vertical straight parts and, therefore, it does not affect the overall oscillation.

1. Will velocity u depend on the distance x along the U-tube?

2. Find an evolution equation for the velocity u in the tube.

3. Find the frequency of the oscillations.

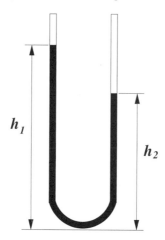

FIGURE 2.3: U-tube.

4. Find the pressure p in the tube as a function of t and x.

2.3.5   Force on a bent garden hose

A garden hose with cross-section area s_1 ends with a nozzle which has a smaller cross-section area s_2. The hose is lying on the ground horizontally with a complicated shape, but the angle between the directions of the water jet entering and leaving the hose is precisely 90 degrees; see figure 2.4. Water enters the hose with velocity u_1 (independent from the distance from the hose axis) under pressure p_1.

In this problem, we will calculate the net force $\mathbf{F}$ exerted on the hose by the water jet. In practice, such a force acts on a bent hose to bend it even more until it curls. An analogous effect appears in plasma moving in a curved magnetic field: in plasma physics, it is often called the garden hose instability.

1. Ignoring viscosity, find velocity u_2 and pressure p_2 of the water jet leaving the nozzle.

2. Find the rate at which the momentum is brought into the hose by the entering jet and the rate at which the momentum is leaving the hose though the nozzle.

3. Compute the net force $\mathbf{F}$ exerted on the hose by the water jet.

2.3.6   Firehose flow

Water from a firehose of internal cross-section area S_1 emerges horizontally through a nozzle of internal cross-section area S_2 at speed u_2; see figure 2.5. In this problem you will need to find the force $\mathbf{F}$ that a fireman needs to exert

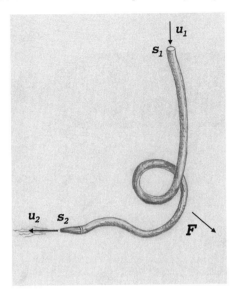

FIGURE 2.4: Bent garden hose.

to hold on to the hose in the *ideal flow approximation* (i.e. ignoring viscosity and compressibility).

1. Formulate Bernoulli's theorem relating the velocities and the pressures at the cross-sections S_1 and S_2, u_1, p_1 and u_2, p_2 respectively; see figure 2.5. Explain why p_2 is equal to the atmospheric pressure p_0.

2. Use the fluid's incompressibility to find the relation between the velocities, u_1 and u_2, and the areas, S_1 and S_2.

3. Find the force **F** needed to hold the firehose. For this, consider the net force acting on the piece of the firehose bounded by cross-sections S_1 and S_2 (including the internal and the external pressure forces) and equate it to the momentum flux due to the fluid volumes crossing S_1 and S_2. **Hint:** Do not forget that the net force $\mathbf{f}_a$ caused by the external atmospheric pressure p_0 has a horizontal component (see figure 2.5).

2.3.7 Shear flow in a strain field

In this problem we will examine the simplest flow configuration with active vortex stretching. Let the velocity field of an ideal (incompressible inviscid) flow consist of two components,

$$\mathbf{u} = \mathbf{u}_s + \mathbf{u}_\sigma,$$

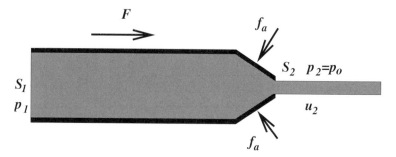

FIGURE 2.5: Water flow in firehose.

with $\mathbf{u}_s$ being a shear flow,

$$\mathbf{u}_s = (sy, 0, 0),$$

where $s = s(t)$ is the shear, a and $\mathbf{u}_\sigma$ being a strain flow,

$$\mathbf{u}_\sigma = (0, -\sigma y, \sigma z),$$

where $\sigma =$ const is the strain.

1. Prove that taken separately, both $\mathbf{u}_s$ (with constant s) and $\mathbf{u}_\sigma$ are solutions to the Navier-Stokes equation and the incompressibility condition.

2. Now consider the flow with $\mathbf{u} = \mathbf{u}_s + \mathbf{u}_\sigma$. Find the vorticity in such a flow. Prove that this flow is a solution of the Navier-Stokes equation and the incompressibility condition. (**Hint:** it is best to consider the vorticity formulation of the Navier-Stokes equation here).

3. Comment on the time dependence of the vorticity and how it arises via the vortex stretching mechanism. Which component of the strain flow makes the vortex lines stretch?

2.3.8 Rankine vortex in a strain field

Consider a vortex with vorticity which has only z-component, $\boldsymbol{\omega} = (0, 0, \omega_z)$, and which is uniformly distributed in a circle of radius R:

$$\omega_z = \Omega = \text{const} \quad \text{for } r^2 = x^2 + y^2 < R, \quad \text{and} \quad \omega_z = 0 \quad \text{for } r^2 \geq R.$$

The flow with such vorticity and with velocity field $\mathbf{u}_R$ which decays at infinity ($\mathbf{u}_R \to 0$ for $r \to \infty$) is called a Rankine vortex; in this case the vortex radius R is constant.

In this example we will consider a case when the velocity field does not decay at infinity. Namely, we will consider a flow with velocity

$$\mathbf{u} = \mathbf{u}_R + \mathbf{u}_\sigma, \tag{2.28}$$

where $\mathbf{u}_R$ is the Rankine vortex and $\mathbf{u}_\sigma$ is a uniform strain field of the form

$$\mathbf{u}_\sigma = (-\sigma x, -\sigma y, 2\sigma z). \tag{2.29}$$

In this case the vortex radius R and the vorticity Ω will be time dependent, $R = R(t), \Omega = \Omega(t)$, because of the vortex stretching produced by the strain.

1. Prove that the Rankine vortex is a solution to the Euler equation for an inviscid fluid. Find the incompressible velocity field $\mathbf{u}_R$ of the Rankine vortex.

2. Prove that the uniform strain field $\mathbf{u}_\sigma$ given by expression (2.29) satisfies the ideal flow equations.

3. Now consider the combination of the Rankine vortex and the strain field as in expression (2.28) and prove that it satisfies the ideal flow equations. Find dependencies $R(t)$ and $\Omega(t)$. Interpret your results in terms of the vortex stretching mechanism.

4. The Burgers vortex is a generalisation of the considered solution to viscous flows. This solution is stationary because the vortex stretching is stabilised by the vorticity diffusion due to viscosity. The stain field in this vortex is the same as in (2.29), but the vorticity profile now is

$$\omega_z = \Omega_0 \, e^{-\lambda r^2},$$

where $\Omega_0 =$const.

Find λ in terms of σ and the viscosity coefficient ν.

2.3.9   Forces produced by a vortex dipole

A vortex dipole consisting of two point vortices with circulations $+\Gamma$ and $-\Gamma$ separated by distance r are injected into the centre of a fluid volume bound by a cylinder of radius r; see figure 2.6. Immediately the vortices start spreading and a boundary layer at the inner walls of the cylinder forms and grows. Consider a moment of time close to the initial one when the vortices are still almost point-like and located near their initial positions, and the boundary layer is thin with thickness $\delta \ll r$.

1. Find the direction and estimate the strength of the viscosity-induced force (per unit length of the cylinder) produced by such a flow onto the retaining cylinder.

2. What is the total momentum of the system (including both the fluid and the cylinder)? Is it changing in time? If initially the cylinder is at rest, will it start moving because of the forces exerted by the flow inside of it? Explain your answers.

3. Estimate the difference of the pressures at the leftmost and the rightmost parts of the flow. Comment on the difference of this result with the one for the inviscid flow (obtained e.g. from Bernoulli's theorem).

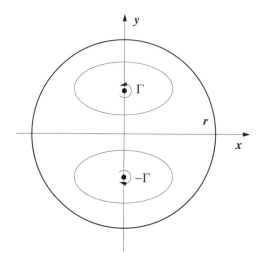

FIGURE 2.6: Vortex pair in a cylindrical container.

2.3.10   Torque produced by a vortex

A point vortex of circulation Γ is injected into the centre of a fluid volume bound by a cylinder of radius R; see figure 2.7.

Immediately the vortex starts spreading and a boundary layer at the inner walls of the cylinder forms and grows. Consider a moment of time close to the initial one when the vortex still almost point-like and the boundary layer is thin and has thickness $\delta \ll R$.

1. Find the boundary layer velocity profile (**Hint:** use the fact that $\delta \ll R$ and reduce your consideration to the one of the boundary layer at a flat plate).

2. Find the torque produced by such a flow onto the retaining cylinder.

2.3.11   Jammed garden hose

From common experience we know that if a garden hose is jammed, i.e. its cross-section is reduced, then the water pressure drops. However, application of Bernoulli's theorem would lead us to conclude that the pressure past the jammed area should recover to the same value as before the jam.

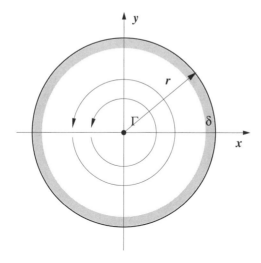

FIGURE 2.7: Vortex in a cylindrical container.

Resolution of this paradox is in realisation that expansion of the pipe cross-section causes flow separation and, therefore, Bernoulli's theorem is not applicable. The present problem will allow us to understand this effect.

A fluid jet emerges from a round pipe of radius r into a wider pipe of radius R; see figure 2.8. The pressure in the narrow pipe is p_1. The sudden expansion results in a flow separation and turbulence in a section of the wider pipe adjacent to the narrow pipe. At some distance downstream, however, the turbulence dies out. For simplicity, we will assume that in both the narrow pipe and in the wide pipe past the turbulent area the velocity profiles are flat, i.e. the velocity is independent of the distance from the pipe axis.

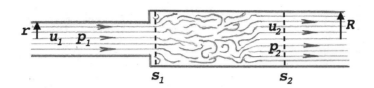

FIGURE 2.8: Sudden change of a pipe cross-section.

1. Use mass conservation and find the relation for the velocity u_2 in the

wide pipe past the turbulent area in terms of the velocity u_1 in the narrow pipe and the radii r and R.

2. Write down the momentum balance equation in a fluid volume bound by a cross-section s_1 in the beginning of the wide pipe (upstream the turbulent area) and a cross-section s_2 in the wide pipe downstream past the turbulent area; see figure 2.8. Neglect the wall friction effects. Find the change of pressure $p_2 - p_1$ between the sections s_2 and s_1 in terms of u_1, r and R. (**Hint:** because of the rapid turbulent mixing, the pressure in the wide pipe at the section s_1 is the same as the narrow pipe pressure p_1.)

3. Now assume that the change in cross-section is gradual and there is no flow separation and turbulence. Use Bernoulli's theorem to find the change of pressure $p_2 - p_1$ in this case. Use your results to comment on the qualitative differences of the pressure behaviour in the systems with the sudden and the gradual changes of cross-sections.

4. Consider a rough model of a jammed garden hose as a pipe which changes its cross-section twice—from large to small and back to large. Will the water pressure in the hose downstream of the jam be greater, equal or less than the initial pressure upstream of the jam in the cases of a strong jam (sudden change of cross-section) and a mild jam (gradual change of cross-section) respectively? Compare these results to the zero-drag result for flows around aerodynamic bodies without flow separation (so-called D'Alembert's paradox) and production of a finite drag in flows around bluff bodies due to the flow separation effect.

2.3.12   Flow through a Borda mouthpiece

Consider a water tank which has an opening in its lower part draining the fluid. It is well-known that the cross-section of the issuing jet of liquid is usually less than that of the opening. For example, for a flat round hole, the contraction coefficient (the ratio of the jet area to the hole area) is known empirically to be approximately 0.62.

A Borda mouthpiece is a reentrant tube in a hydraulic reservoir; see figure 2.9. Its contraction coefficient can be calculated more simply than for other discharge openings. Let the diameter of the Borda mouthpiece be d_m and the diameter of the jet after it has reached the maximal contraction (place called *vena contracta*) be d_j. By definition, the contraction coefficient is the ratio of the respective cross-section areas $s_j/s_m = d_j^2/d_m^2$.

1. Let the fluid pressure in the reservoir near the entrance into the Borda mouthpiece be p_1 and the pressure outside of the reservoir be atmospheric, p_0. Find the velocity of the jet u at the *vena contracta*.

2. Write down the momentum balance relation over the entire volume of

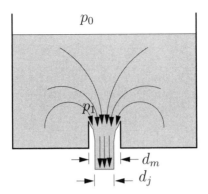

FIGURE 2.9: Flow through a Borda mouthpiece.

fluid in the reservoir (including the Borda mouthpiece). (**Hint**: you need to consider the force imbalance with respect to the equilibrium situation when the Borda mouthpiece is locked by a solid surface at its exit). Substitute your expression for the jet velocity into the momentum balance relation and find the contraction coefficient.

2.3.13 Water barrel on wheels

A cylindrical barrel of diameter D and mass m (when empty) is filled with water to an initial height h_0. The barrel is placed on wheels, and at its side, right at the bottom, it has a hole discharging a water jet of diameter d; see figure 2.10. The water jet produces a reaction force acting on the barrel, and the latter starts moving.

1. Find the velocity u at which the water jet is leaving the barrel through the hole if the water is filled to level h. (**Hint**: strictly speaking, Bernoulli's law should be used in the reference frame moving with the barrel. However, you should assume that the barrel speed is much less than u and, therefore, can be neglected when calculating u).

2. Using the water volume conservation, calculate the rate of change of the water level in time, $\dot{h}(t)$, in terms of the velocity u.

3. Use your answers to questions 1 and 2 to find the dependence of the jet velocity on time, $u(t)$.

4. Find the total momentum M which has left the barrel with the water jet from the time the barrel was full to the moment it was empty.

5. Find the terminal velocity V of the empty barrel assuming that its motion on the wheels is frictionless.

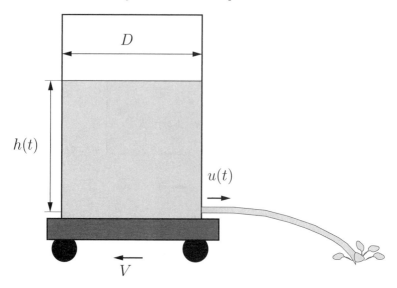

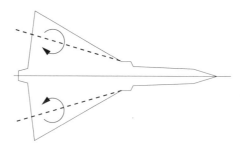

FIGURE 2.10: Flow from a moving barrel.

2.3.14 Vortex lift

Some aeroplane wings have sharply swept leading edges, which generates vortices on the the upper sides of both wings; see figure 2.11. Examples include a delta winged F-106 military jet and a commercial one (no longer in use) — the Concorde. Each vortex is trapped by the airflow and remains fixed to the upper surface of the wing. The major advantage of vortex lift is that it allows angles of attack that would stall a normal wing. The vortices also produce high drag which can help to slow down the aircraft. This is why the vortex lift is used during (high angle of attack) landing of most supersonic jets.

FIGURE 2.11: Vortices over sharply swept plane wings.

In this problem we will aim to understand how vortices produce lift. For this, we will consider a simplified situation in which an infinite straight vortex with circulation Γ is placed parallel to an infinite flat plate (an idealised wing);

see figure 2.12. For simplicity, we will assume that the flow is inviscid and incompressible.

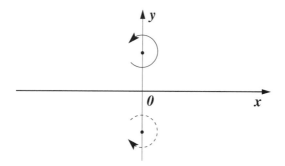

FIGURE 2.12: Vortex over an infinite flat plate located at $y = 0$. The image vortex is below the plate (dashed line).

1. Formulate the free-slip boundary conditions on the plate.

2. Find the velocity field produced by the vortex of circulation Γ on the top surface of the plate. (**Hint**: use the vortex image method.)

3. Find the pressure distribution on the top surface of the plate assuming that the surrounding pressure (i.e. far away from the vortex) is atmospheric, p_0.

4. Assuming that the pressure at the lower side of the plate is uniform and equal to the atmospheric value, find the total force on the plate per unit length in the vortex direction.

2.3.15 Water clock

A water clock is an axisymmetric vessel with a small exit hole of radius a in the bottom; see the figure 2.13. Find the vessel shape for which the water level falls equal heights in equal intervals of time. (**Hint:** the hole is so small that the water passes through it very slowly and its velocity can be found from Bernoulli's theorem for stationary flows.)

2.3.16 Reservoir with regulated water level

A rectangular sluice gate is fitted at the base of a reservoir wall with a pivot in the arrangement shown in the figure 2.14. The gate is designed to regulate the level of water in the reservoir by opening when the water level to the right, h, reaches a certain depth.

1. Find the pressure as a function of the coordinate.

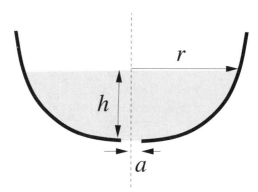

FIGURE 2.13: Water clock.

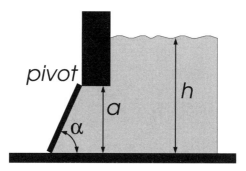

FIGURE 2.14: Reservoir with a sluice gate.

2. The gate's width is b and it is made out of steel of uniform thickness. It is positioned at α degrees to the ground and the vertical distance from the pivot to the ground is a. Determine the mass of the gate M if a water level h will just cause the gate to open. (**Hint:** the torque produced at the pivot by the gate's weight is balanced by the torque due the water pressure.)

2.3.17   Energy of ideal irrotational flows

1. Ideal fluid moves irrotationally in a simply connected region V bounded by a closed surface S, so that $\mathbf{u} = \nabla \phi$, where ϕ is the velocity potential. Show that the kinetic energy

$$E = \frac{1}{2} \int_V \mathbf{u}^2 \, d\mathbf{x}$$

can be written in form

$$E = \frac{1}{2} \int_S \phi \, (\mathbf{n} \cdot \nabla \phi) \, dS,$$

where $\mathbf{n}$ denotes a unit vector normal to the surface element dS.

2. Ideal fluid moves in a bounded simply connected region V, and the normal component of velocity $\mathbf{u} \cdot \mathbf{n}$ is given at each point on the boundary of V. Show that there is at most one irrotational flow in V which satisfies this boundary condition.

3. Ideal fluid occupies the gap $r_1 < r < r_2$ between two infinitely long cylinders, which are fixed. The irrotational flow between them is

$$\mathbf{u} = \frac{\Gamma}{2\pi r} \hat{\boldsymbol{\theta}},$$

where $\Gamma = $ const and $\hat{\boldsymbol{\theta}}$ is a unit vector in the azimuthal direction. As there is no normal velocity on either bounding surface, $r = r_1$ or $r = r_2$, could one use the result of part 1 to concluded that the kinetic energy is zero? Explain your answer.

2.4 Solutions

2.4.1 Model solution to question 2.3.1

1. The incompressibility condition in terms of the velocity field is $\nabla \cdot \mathbf{u} = 0$. The vorticity field satisfies the same equation, $\nabla \cdot \boldsymbol{\omega} = 0$, by virtue of its definition $\boldsymbol{\omega} = \nabla \times \mathbf{u}$: because the divergence of any curl is zero, $\nabla \cdot (\nabla \times \mathbf{u}) = 0$.

2. Using the definition of the material time derivative, the conservation of the field s along the trajectories of fluid particles is

$$D_t s \equiv \partial_t s + \mathbf{u} \cdot \nabla s = 0. \tag{2.30}$$

3. Taking the curl of the Euler equation (2.27), we have the vorticity evolution equation:

$$\partial_t \boldsymbol{\omega} = \nabla \times (\mathbf{u} \times \boldsymbol{\omega}) + (\nabla \rho \times \nabla p)/\rho^2. \tag{2.31}$$

4. If the passive scalar $s(\mathbf{x}, t)$ depends on $\mathbf{x}$ and t only via p and ρ, $s = s(p, \rho)$

$$\nabla s \cdot (\nabla \rho \times \nabla p) = [(\partial_\rho s)\nabla \rho + (\partial_p s)\nabla p] \cdot (\nabla \rho \times \nabla p) = 0$$

for both barotropic and baroclinic fluids.

5. Taking the partial time derivative of $\lambda = \nabla s \cdot \boldsymbol{\omega}$ and using the equations (2.31) and (2.30) and the vector identity (2.26), as well as the identity $\nabla s \cdot (\nabla \rho \times \nabla p) = 0$, we have:

$$\partial_t (\nabla s \cdot \boldsymbol{\omega}) = (\nabla \partial_t s) \cdot \boldsymbol{\omega} + \nabla s \cdot \partial_t \boldsymbol{\omega} = \nabla s \cdot (\nabla \times (\mathbf{u} \times \boldsymbol{\omega})) - \boldsymbol{\omega} \cdot \nabla (\mathbf{u} \cdot \nabla s) =$$

$$-\mathbf{u} \cdot \nabla (\boldsymbol{\omega} \cdot \nabla s).$$

Thus,

$$D_t \lambda \equiv \partial_t \lambda + \mathbf{u} \cdot \nabla \lambda = 0. \tag{2.32}$$

2.4.2 Model solution to question 2.3.2

1. The pressure in this part of the jet remains atmospheric, p_0. Thus, using Bernoulli's theorem we have

$$\frac{u^2}{2} = \frac{u_0^2}{2} + gz.$$

Therefore, $u = \sqrt{u_0^2 + 2gz}$.

From the mass conservation we have $ud^2 = u_0 d_0^2$. So, $d = d_0 \sqrt{u_0/u} = d_0 (1 + 2gz/u_0^2)^{-1/4}$.

2. At the impact point with the solid surface at the centre of the jet the fluid velocity is zero. Thus,

$$p = p_0 + \frac{\rho u_0^2}{2} + \rho g h.$$

2.4.3 Model solution to question 2.3.3

1. Because u is independent of the distance from the centre of the pipe, it must also be independent from the distance along the pipe x by incompressibility. Thus, u is a function of time only, $u = u(t)$.

2. The terminal velocity can be found from the Bernoulli's theorem, which gives $u_\infty = \sqrt{2gh}$.

3. Now, we are dealing with a non-stationary flow and, therefore, we need to use the time-dependent version of the Bernoulli's theorem. Taking into account that for the uniform velocity directed along x the velocity potential is $\phi = ux$, and that the pressure at the far end of the pipe is atmospheric (p_0), we have

$$\dot{u} l + \frac{u^2}{2} + \frac{p_0}{\rho} = C, \tag{2.33}$$

where C is a constant.

On the other hand, the pressure at the surface of the flume is also atmospheric, so

$$\frac{p_0}{\rho} + gh = C,\tag{2.34}$$

where C is the same constant. Here, we neglected the flow velocity at the flume surface because the flume surface area is much greater than the cross-section area of the pipe.

Combining (2.33) and (2.34), we have

$$\dot{u}l + \frac{u^2}{2} = gh = \frac{u_\infty^2}{2}.\tag{2.35}$$

Integrating this equation, we have

$$u = u_\infty \tanh \frac{u_\infty t}{2l}.\tag{2.36}$$

We see that initially the velocity grows almost linearly, $u \approx \frac{u_\infty^2 t}{2l}$, and it approaches its asymptotic value u_∞ for large time.

4. At an arbitrary (not too small) distance x along the pipe, we have

$$\dot{u}x + \frac{u^2}{2} + \frac{p - p_0}{\rho} = gh.\tag{2.37}$$

Solving for p and substituting u from (2.36) one can get the required expression for the pressure:

$$p = p_0 + \frac{\rho}{2}(1 - x/l)(u_\infty^2 - u^2) = p_0 + \frac{\rho u_\infty^2}{2}\left(1 - \frac{x}{l}\right)\left[1 - \left(\tanh \frac{u_\infty t}{2l}\right)^2\right].$$

2.4.4   Model solution to question 2.3.4

1. Because u is independent of the distance from the centre of the tube, it must also be independent from the distance along the pipe x by incompressibility. Thus, u is a function of time only, $u = u(t)$.

2. Since we are dealing with a non-stationary flow, we need to use the time-dependent version of Bernoulli's theorem. Let us define x distance along the tube starting from its top left end, i.e. x will first trace the left part of the tube downwards until $x = h$ (h being the height of the U-tube) and then the right part upwards. Taking into account that for the uniform velocity directed along x the velocity potential is $\phi = ux$, we have

$$\dot{u}x + \frac{u^2}{2} + \frac{p}{\rho} + G = C(t),\tag{2.38}$$

where $C(t)$ is a function of time (to be found later) and $G = g(h - x)$ is the gravitational potential.

3. Taking the difference of this equation at the two ends of the tube and taking into account that the pressure at the ends of the tube is atmospheric (p_0), we have

$$\ddot{u}l - g(h_1 - h_2) = 0, \tag{2.39}$$

where $l = h_1 + h_2 = h_1(0) + h_2(0) = \text{const}$ is the total length occupied by the fluid. Taking into account that $u = \dot{h}_2 = -\dot{h}_1$ we have for $s = h_2 - h_1$:

$$\ddot{s}l + 2gs = 0. \tag{2.40}$$

Solving this equation we have

$$s(t) = A\cos(\omega t), \tag{2.41}$$

where $A = h_2(0) - h_1(0)$ is the oscillation amplitude and the oscillation frequency is given by

$$\omega = \sqrt{\frac{2g}{l}}. \tag{2.42}$$

Interestingly, this expression coincides with the frequency formula for a pendulum which consists of a point mass suspended on a massless string of length $l/2$.

4. Differentiating expression (2.41), we get the following solution for the velocity

$$u(t) = U\sin(\omega t), \tag{2.43}$$

where $U = \frac{\omega}{2}[h_1(0) - h_2(0)]$. Note that the oscillation phase is chosen in (2.41) and (2.43) to ensure that the initial condition $u(0) = 0$ is satisfied. Similarly, for h_1 we have

$$h_1(t) = (h_1(0) - l/2)\cos(\omega t) + l/2, \tag{2.44}$$

From (2.38), we find for the pressure:

$$\frac{p}{\rho} = -U\omega(x - h_1(t))\cos(\omega t) + g(x - h_1(t)) + \frac{p_0}{\rho}, \tag{2.45}$$

where $C(t)$ is chosen to satisfy condition $p = p_0$ at $x = h_1(t)$.

2.4.5　Model solution to question 2.3.5

1. From mass conservation,

$$u_2 = \frac{u_1 s_1}{s_2}. \tag{2.46}$$

From Bernoulli's theorem we have

$$p_2 - p_1 = \frac{\rho}{2}(u_1^2 - u_2^2) = \frac{\rho}{2}u_1^2\left(1 - \frac{s_1^2}{s_2^2}\right). \tag{2.47}$$

2. The rate at which the momentum is brought into the hose is $\rho s_1 u_1^2\,\hat{\mathbf{x}}$. The rate at which the momentum is leaving the hose is $\rho s_2 u_2^2\,\hat{\mathbf{y}}$. Here, $\hat{\mathbf{x}}$ and $\hat{\mathbf{y}}$ are the unit vectors along the x- and y-axes respectively.

3. Consider the entire volume of fluid in the hose. The total value of momentum entering this volume per unit time must be compensated by the net force acting onto this fluid volume. In turn, this net force is equal to the pressure force through the cross-sections on both ends of the hose, $p_1 s_1\,\hat{\mathbf{x}} - p_2 s_2\,\hat{\mathbf{y}}$, plus the net reaction force exerted by the hose walls onto the fluid $\mathbf{R}$. The net force exerted onto the garden hose by the jet is $\mathbf{F} = -\mathbf{R}$. So,

$$\mathbf{F} = \rho s_1 u_1^2\,\hat{\mathbf{x}} - \rho s_2 u_2^2\,\hat{\mathbf{y}} + p_1 s_1\,\hat{\mathbf{x}} - p_2 s_2\,\hat{\mathbf{y}}. \tag{2.48}$$

Substituting u_2 and p_2 from equations (2.46) and (2.47) respectively, we have

$$\mathbf{F} = (\rho u_1^2 + p_1)s_1\,\hat{\mathbf{x}} - \left[p_1 + \frac{\rho}{2}u_1^2\left(1 + \frac{s_1^2}{s_2^2}\right) \right] s_2\hat{\mathbf{y}}. \tag{2.49}$$

2.4.6   Model solution to question 2.3.6

1. Bernoulli's theorem states: $\frac{u_1^2}{2} + p_1/\rho = \frac{u_2^2}{2} + p_2/\rho$. At the nozzle, water is in contact with the atmosphere (via the free surface). Thus $p_2 = p_0$.

2. The incompressibility means that the water volume flux is the same at all cross-sections, so $u_1 S_1 = u_2 S_2$.

3. The net force on the piece of the firehose bounded by cross-sections S_1 and S_2 is equal to the force F applied by the fireman plus the force applied by the pressure p_1 at S_1 plus the force applied by the atmospheric pressure (acting in the horizontal direction through the jet and the outer surface of the convergent nozzle; see $\mathbf{f}_a$ in figure 2.5) must be equal to the difference in the momentum flux through the sections:

$$F - p_0 S_1 + p_1 S_1 = \rho(u_2^2 S_2 - u_1^2 S_1),$$

or, taking into account Bernoulli's theorem and the mass conservation:

$$F = -\frac{\rho u_2^2}{2 S_1}(S_1 - S_2)^2.$$

Surprised that the force is negative? See a discussion in [29].

2.4.7   Model solution to question 2.3.7

1. First of all, the incompressibility condition $\nabla \cdot \mathbf{u}$ is satisfied since

$$\nabla \cdot \mathbf{u}_s = \partial_x(sy) = 0$$

and

$$\nabla \cdot \mathbf{u}_\sigma = -\partial_y(\sigma y) + \partial_z(\sigma z) = 0.$$

Both $\mathbf{u}_s$ and $\mathbf{u}_\sigma$ are linear functions of $\mathbf{x}$ and, therefore, the viscous term is zero. Since s and σ are constant, $\partial_t \mathbf{u}_s = 0$ and $\partial_t \mathbf{u}_\sigma = 0$.

Now,

$$(\mathbf{u}_s \cdot \nabla)\mathbf{u}_s = sy\, \partial_x(sy)\, \hat{\mathbf{x}} = 0$$

and

$$(\mathbf{u}_\sigma \cdot \nabla)\mathbf{u}_\sigma = \sigma y\, \partial_y(\sigma y)\, \hat{\mathbf{y}} + \sigma z\, \partial_z(\sigma z)\, \hat{\mathbf{z}} = \sigma^2(y\, \hat{\mathbf{y}} + z\, \hat{\mathbf{z}}).$$

Thus, $\mathbf{u}_s$ solves the Navier-Stokes equation with pressure $p =$const and $\mathbf{u}_\sigma$ solves the Navier-Stokes equation with pressure

$$p = \text{const} - \frac{1}{2}\rho\sigma^2(y^2 + z^2).$$

2. For the vorticity we have:

$$\boldsymbol{\omega} = \nabla \times \mathbf{u} = \nabla \times \mathbf{u}_s + \nabla \times \mathbf{u}_\sigma = s\, \hat{\mathbf{z}}$$

(because $\nabla \times \mathbf{u}_\sigma = 0$).

Since there is only one non-zero component of the vorticity, $\boldsymbol{\omega} = (0, 0, \omega)$, ($\omega = s$), we will only need to consider the z-component of the vorticity equation. Because ω is uniform in space, the viscous term is zero. For the same reason, $\mathbf{u} \cdot \nabla\omega = 0$.

For the vortex stretching term we have

$$(\boldsymbol{\omega} \cdot \nabla)\mathbf{u} = s\, \partial_z(\sigma z\, \hat{\mathbf{z}}) = s\sigma\, \hat{\mathbf{z}}.$$

Thus, the vorticity equation in our case is

$$\partial_t s = s\sigma,$$

which has the following solution

$$s(t) = s_0\, e^{\sigma t},$$

where s_0 is the initial value of the shear at $t = 0$.

3. As we can see, the vorticity is growing exponentially in time, $\omega(t) = s(t) = s_0\, e^{\sigma t}$. This is due to the z-component of the strain flow which pulls the fluid particles apart in the z-direction exponentially in time. At the same time, due to volume conservation, the fluid particles are pushed closer to each other in the y-direction making the density of the vortex lines greater—hence the vorticity amplification.

2.4.8 Model solution to question 2.3.8

1. To prove that the Rankine vortex is a solution to the Euler equation, we will use the vorticity form of this equation, (2.8) with $\nu = 0$. Since the vorticity has only the z-component, we will need to consider only the z-component of the vorticity equation (the other two components have zeroes on both sides):

$$(\partial_t + (\mathbf{u} \cdot \nabla))\omega_z = (\boldsymbol{\omega} \cdot \nabla))u_z. \tag{2.50}$$

In fact, not only the Rankine vortex but also any vortex with an arbitrary axisymmetric distribution of vorticity $\omega_z(r)$ is a solution. Indeed, from the incompressibility, we have that there is no radial velocity component: hence $(\mathbf{u} \cdot \nabla)\omega_z(r) = 0$. From stationarity $\partial_t \omega_z(r) = 0$, and because $u_z = 0$ the vortex stretching term (i.e. the right-hand side of (2.50)) is also zero. Thus we see that equation (2.50) is satisfied for an arbitrary axisymmetric distribution of vorticity $\omega_z(r)$ including the Rankine vortex as a special case.

The velocity field $\mathbf{u}_R$ of the Rankine vortex can be found using the Stokes theorem,

$$\oint_{C_R} \mathbf{u}_R \cdot d\boldsymbol{\ell} = \int_{S_R} \omega_z \, dx dy,$$

where C_R is a circular contour of radius R centred at $\mathbf{x} = 0$ and S_R is the surface spanning this contour.

This gives

$$\mathbf{u}_R = \frac{\Omega}{2} r \hat{\boldsymbol{\theta}} \quad \text{for } r < R \quad \text{and} \quad \mathbf{u}_R = \frac{\Omega R^2}{2r} \hat{\boldsymbol{\theta}} \quad \text{for } r \geq R,$$

where $\hat{\boldsymbol{\theta}}$ is the unit vector in the azimuthal direction.

2. The uniform strain field $\mathbf{u}_\sigma$ given by the expression (2.29) is a potential flow with velocity potential

$$\phi = \frac{\sigma}{2}(-r^2 + 2z^2), \tag{2.51}$$

which can be verified by substituting into $\mathbf{u}_\sigma = \nabla \phi$ and confirming that it is satisfied.

For the potential flow the fluid equation reduces to the Laplace equation,

$$\nabla^2 \phi = 0,$$

which is obviously satisfied for the potential (2.51).

3. Consider the combination of the Rankine vortex and the strain field as in expression (2.28). We can still use the vorticity equation (2.50) but now

with non-zero stretching term. The vortex advection term (the second term on the left-hand side) is zero both inside and the outside of the Rankine vortex because in both places $\nabla \omega_z = 0$. Outside of the vortex all terms in the vorticity equation are zero and it is obviously satisfied. Inside the vortex we have

$$\partial_t \Omega = \Omega \partial_z \mathbf{u}_z = 2\Omega\sigma, \tag{2.52}$$

so

$$\Omega = \Omega(0)e^{2\sigma t}.$$

Dependence $R(t)$ can be obtained from the Kelvin's theorem for conservation of the circulation. Conservation of the circulation over C_R, the circular contour of radius R, means $\Omega R^2 = $ const, i.e.

$$R = R(0)e^{-\sigma t}.$$

The vorticity amplification is produced by the vortex stretching mechanism activated by the z-component of the strain field.

4. Let us substitute the Burgers vortex, $\omega_z = \Omega_0\, e^{-\lambda r^2}$, into a steady-state version of the z-component of the vorticity equation (2.8), but now with finite ν.

$$(\mathbf{u}_\sigma \cdot \nabla)e^{-\lambda r^2} = e^{-\lambda r^2}\partial_z u_{\sigma z} + \nu\nabla^2 e^{-\lambda r^2}. \tag{2.53}$$

Substituting for $\mathbf{u}_\sigma$ and differentiating, we have

$$-\sigma(-2\lambda r^2)e^{-\lambda r^2} = 2\sigma e^{-\lambda r^2} + \nu(-4\lambda + 4\lambda^2 r^2)e^{-\lambda r^2}. \tag{2.54}$$

This equation is satisfied if

$$\lambda = \frac{\sigma}{2\nu}.$$

2.4.9 Model solution to question 2.3.9

1. From figure 2.6 it is clear that the viscosity induces a force along the x-axis in the negative direction. This is because the negative x-momentum will be transferred (in equal proportions) to the top and at the bottom parts of the cylinder via the viscous stress in the boundary layer. There will be no y-component of the force because, due to symmetry, the y-momentum transfers at the top and at the bottom parts will sum up to zero.

The velocity field produced by the vortex dipole at the walls can be estimated as $u \sim \Gamma/r$. Correspondingly, the viscous stress is $\sigma_{xy} \sim \rho\nu u/\delta \sim \rho\nu\Gamma/(\delta r)$. This gives the following estimate for the viscous force on the cylinder per unit length in the z-direction,

$$\mathbf{F} = (F_x, 0) \sim (-2\pi\rho\nu\Gamma/\delta, 0).$$

Notice that this force is independent of the radius of the cylinder r!

2. Assuming that there is no external force on the cylinder (i.e. through its outer surface), the total momentum of the cylinder and the fluid without it is conserved by Newton's second law. But by symmetry, the total initial momentum of the fluid flow is zero. Thus the cylinder which is initially at rest will remain at rest for infinite time.

3. Since the net friction force on the inner walls of the cylinder is finite, and because the total force must be zero (the total momentum is conserved) there must arise a difference of the pressures at the left and the right inner surfaces of the cylinder to compensate the friction force. One can use, for example, formula (2.18) in which $g = 0$. This gives the following estimate for the pressure difference,

$$p_{\text{left}} - p_{\text{right}} \sim \frac{\rho\nu\Gamma}{\delta r}.$$

For the inviscid flow we would get from Bernoulli's theorem $p_{\text{left}} - p_{\text{right}} = 0$ because both the leftmost and the rightmost points are the stagnation points where the velocity is zero. This agrees with the limit $\nu \to 0$ of the above expression.

2.4.10   Model solution to question 2.3.10

1. Let us use the fact that $\delta \ll R$ and reduce our consideration to the one of the boundary layer at a flat plate by ignoring the boundary's curvature. Let us introduce a local Cartesian coordinate frame with the origin at the inner surface of the cylinder, the x-axis tangent to the surface and the y-axis normal to it.

   In this geometry, the velocity will only have the x-component which depends on y and t,
   $$\mathbf{u} = (u(y, t), 0),$$
   there will be no pressure gradient ($\partial_x p = 0$ because all points on the boundary are equivalent and $\partial_y p = 0$ because there is no motion in the y-direction), and $(\mathbf{u} \cdot \nabla)\mathbf{u} = 0$.

   Thus, we have the following equation for the x-component of the velocity,
   $$\partial_t u = \nu\partial_{yy} u, \tag{2.55}$$
   which is just the heat equation.

   The no-slip boundary condition to be imposed on the solution of this equation is
   $$u(0, t) = 0. \tag{2.56}$$
   Also, because the boundary layer is thin, the velocity field produced by the vortex,
   $$\mathbf{u} = (u_\infty, 0) = \left(\frac{\Gamma}{2\pi R}, 0\right),$$

can be viewed as a boundary condition at $y \to +\infty$:

$$u(+\infty, t) = \frac{\Gamma}{2\pi R}. \tag{2.57}$$

On the other hand, because there is no boundary layer initially, we have the following initial condition,

$$u(y, 0) = u_\infty = \frac{\Gamma}{2\pi R}. \tag{2.58}$$

We will seek a self-similar solution,

$$u(y, t) = u_\infty f(\eta),$$

with the similarity variable η given by

$$\eta = \frac{y}{\sqrt{\nu t}},$$

and the boundary conditions

$$f(0) = 0 \quad \text{and} \quad f(+\infty) = 1. \tag{2.59}$$

Note that the initial condition for u at $t = 0$ coincides with the second boundary condition for f, i.e. $f(+\infty) = 1$.

Applying the chain rule for differentiation, we have

$$\partial_t u = u_\infty f' \partial_t \eta = -u_\infty f' \frac{y}{2\nu^{1/2} t^{3/2}},$$

and

$$\partial_{yy} u = u_\infty \left(f''(\partial_y \eta)^2 + f' \partial_{yy} \eta \right) = u_\infty f'' \frac{1}{\nu t},$$

so that the equation for f becomes

$$f'' = -\frac{1}{2}\eta f'.$$

Integrating this equation, we get

$$f' = C_1 e^{-\eta^2/4},$$

and integrating again, we arrive at

$$f = C_1 \int_0^\eta e^{-s^2/4}\, ds + C_2,$$

where C_1 and C_2 are constants which have to be found to match the boundary conditions. Taking into account that $\int_0^\eta e^{-s^2/4}\, ds = \sqrt{\pi}$, we have

$$C_1 = \frac{1}{\sqrt{\pi}} \quad \text{and} \quad C_2 = 0.$$

For the velocity profile we have

$$u = \frac{u_\infty}{\sqrt{\pi}} \int_0^{y/\sqrt{\nu t}} e^{-s^2/4}\, ds.$$

2. The shear stress at the boundary is

$$\sigma_{xy} = \rho\nu\partial_y u|_{y=0} = \frac{\rho\nu u_\infty}{\sqrt{\pi\nu t}} e^{-(y/\sqrt{\nu t})^2/4}|_{y=0} = \frac{\rho\nu u_\infty}{\sqrt{\pi\nu t}}.$$

Therefore the torque produced onto the retaining cylinder (per unit length of the cylinder) is

$$\tau = R(2\pi R)\frac{\rho\nu u_\infty}{\sqrt{\pi\nu t}} = \rho\Gamma R\sqrt{\frac{\nu}{\pi t}}.$$

2.4.11 Model solution to question 2.3.11

1. From mass conservation we find

$$u_2 = \frac{u_1 r^2}{R^2}. \tag{2.60}$$

2. Momentum influx into the volume bound by the sections s_1 and s_2 occurs though the narrow pipe at a rate $\pi r^2 \rho u_1^2$. The momentum outflux from the same volume occurs through the section s_2 at a rate $\pi R^2 \rho u_2^2$. Thus the net rate of the momentum influx into the fluid volume bounded by s_1 and s_2 and the pipe walls is $\pi r^2 \rho u_1^2 - \pi R^2 \rho u_2^2$; it has to be balanced by net force acting on this volume, namely by the difference of the pressure forces at s_1 and s_2:

$$(p_2 - p_1)\pi R^2 = \pi\rho(r^2 u_1^2 - R^2 u_2^2) = \pi\rho u_1^2 r^2\left(1 - \frac{r^2}{R^2}\right), \tag{2.61}$$

or

$$p_2 - p_1 = \rho u_1^2 \frac{r^2}{R^2}\left(1 - \frac{r^2}{R^2}\right). \tag{2.62}$$

3. For the gradual expansion case, from Bernoulli's theorem we have

$$p_2 - p_1 = \frac{\rho}{2}(u_1^2 - u_2^2) = \frac{\rho}{2}u_1^2\left(1 - \frac{r^4}{R^4}\right), \tag{2.63}$$

which is greater than the sudden-expansion pressure change (2.62) by $\frac{\rho}{2}u_1^2\left(1 - \frac{r^2}{R^2}\right)^2$.

4. In the mildly jammed hose, the flow pattern will be fully symmetric with respect to the middle point of the jammed area, and the pressure downstream of the jam will have exactly the same value as the upstream pressure. Thus, there will be no net force exerted by the flow onto the jammed area.

   In the strongly jammed hose, one can assume that contraction from large to small cross-section occurs without a flow separation. Thus the pressure difference between the upstream and the jammed parts will be the same as in the mild jam case and could be found from Bernoulli's theorem. However, the downstream flow will be separated and the pressure will be less that the one predicted by Bernoulli's theorem, as we established in the previous part of this problem. Thus, there will be a net force onto the jammed area in the direction along the flow.

There is a direct analogy with flows past solid bodies. Namely, if there is no flow separation (i.e. the body shape is aerodynamic) there will be no drag force—result known as d'Alembert's paradox. On the other hand, if there is separation (bluff body shape), then the pressure behind the body will not recover to the levels of the upstream flow as would be prescribed by Bernoulli's theorem. This disbalance of the upstream and the downstream pressures would cause a finite drag.

2.4.12   Model solution to question 2.3.12

1. By Bernoulli's theorem,

$$p_1 - p_0 = \frac{\rho}{2}u^2. \tag{2.64}$$

2. Force imbalance with respect to the equilibrium situation when the Borda mouthpiece is locked by a solid surface is equal to the (removed) reaction force of such a solid surface, i.e. $(p_1 - p_0)s_m$. It has to be equal to the rate at which momentum is carried away by the jet, i.e. $\rho u^2 s_j$:

$$(p_1 - p_0)s_m = \rho u^2 s_j. \tag{2.65}$$

Combining (2.64) and (2.65), we finally have

$$s_j/s_m = 0.5.$$

2.4.13   Model solution to question 2.3.13

1. The pressure at the water surface and the one at the jet are both equal to the atmospheric pressure. Thus, ignoring the fluid velocity at the surface $(\dot{h})$ and the barrel velocity, from Bernoulli's theorem we have

$$\frac{u^2}{2} = gh. \tag{2.66}$$

2. From mass conservation we have $\dot{h} = -u(d/D)^2$.

3. Substituting this formula into the time-differentiated equation (2.66), we have: $u\dot{u} = g\dot{h} = -ug(d/D)^2$. Integrating, we get:

$$u = \sqrt{2gh_0} - g(d/D)^2\, t. \tag{2.67}$$

4. The momentum M which has left the barrel with the water jet from the time $t = 0$ when the barrel was full to the moment $t = t_*$ when it was empty is

$$M = \int_0^{t_*} (\rho u)(\pi d^2) u\, dt.$$

Substituting in here from (2.67) and integrating, we have $M = \frac{2\pi \rho D^2}{3}\sqrt{2gh_0^3}$. Here, we took into account that $u(t_*) = 0$.

Note that d is usually smaller than the hole diameter due to a transient jet contraction right near the hole. However, the jet diameter d drops out of the answer for M. Can you explain why?

5. The total momentum is conserved. Therefore, the total momentum carried away by the water jet will be equal in magnitude and opposite in sign to the momentum acquired by the empty barrel, $mV = M$. So,

$$V = -\frac{2\pi \rho D^2}{3m}\sqrt{2gh_0^3}.$$

2.4.14 Model solution to question 2.3.14

1. The free-slip boundary condition means that the tangential velocity is not imposed and only the non-penetration condition is enforced, i.e. at the plate there is no normal component, $u_y = 0$ at $y = 0$.

2. Suppose the vortex of circulation Γ is at distance d from the plate, i.e. at $(x, y) = (0, d)$. To satisfy the free-slip boundary condition, the image vortex of circulation $-\Gamma$ has to be placed at $(x, y) = (0, -d)$. Then at $y = 0$ we have $u_y = 0$ and the x-component of velocity at $y = 0$ is the double x-velocity produced by the vortex Γ, i.e.

$$u_x(x, 0) = \frac{2\Gamma d}{2\pi(x^2 + d^2)}.$$

3. The pressure is to be calculated from Bernoulli's theorem,

$$p(x, 0) = p_0 - \frac{\rho[u_x(x,0)]^2}{2} = p_0 - \frac{\Gamma^2 d^2}{\pi^2(x^2 + d^2)^2}.$$

4. Thus the force on the plate (per unit length in the vortex direction) is

$$F = \int_{-\infty}^{\infty} \frac{\Gamma^2 d^2}{\pi^2(x^2 + d^2)^2}\, dx.$$

This integral can be computed as

$$-\frac{\Gamma^2 d^2}{\pi^2}\frac{\partial}{\partial(d^2)}\int_{-\infty}^{\infty}\frac{1}{(x^2 + d^2)}\, dx = -\frac{\Gamma^2 d^2}{\pi^2}\frac{\partial}{\partial(d^2)}\left[\frac{1}{d}\arctan(x/d)\right]_{-\infty}^{\infty} = \frac{\Gamma^2}{2\pi d}.$$

2.4.15   Model solution to question 2.3.15

Since the pressures at the water surface and at the bottom hole are the same, from Bernoulli's theorem for stationary flows we have $gh = u^2/2$ i.e. $u = \sqrt{2gh}$; see figure 2.13.

From the water volume conservation we have:

$$\dot{h}r^2 = -ua^2.$$

Thus,

$$\dot{h}r^2 = -a^2\sqrt{2gh},$$

and from the condition that the water level is reduced at a constant rate, $\dot{h} = C = \text{const}$, we have

$$h(r) = \frac{C^2}{2ga^4}\, r^4.$$

2.4.16   Model solution to question 2.3.16

1. An expression for the pressure as a function of the coordinate can be found from the Bernoulli's theorem applied for the case when the velocity is zero:

$$p = p_0 + \rho g z,$$

there ρ is the water density, p_0 is the atmospheric pressure, g is the gravity constant and z is the depth measured from the water surface.

2. The torque produced on the gate by its own gravity is

$$Mg\frac{a}{2}\cos\alpha.$$

Since the pressure forces are normal to the gate, their torque at the pivot is:

$$\int_0^{a/\sin\alpha} (p - p_0)bs\, ds = b\int_0^{a/\sin\alpha} \rho g(h - a + s\sin\alpha)s\, ds =$$

$$b\rho g\left[(h - a)\frac{a^2}{2\sin^2\alpha} + \frac{a^3}{3\sin^2\alpha}\right].$$

Thus,

$$M = \frac{ba\rho}{\sin^2\alpha\cos\alpha}\left[h - \frac{a}{3}\right].$$

2.4.17 Model solution to question 2.3.17

1. Integrating by parts we can write the kinetic energy as:

$$E = \frac{1}{2} \int_V (\nabla \phi)^2 \, d\mathbf{x} = \frac{1}{2} \int_V [\nabla \cdot (\phi \nabla \phi) - \phi \nabla^2 \phi] \, d\mathbf{x}.$$

The second term here is zero by the incompressibility condition, $\nabla \cdot \mathbf{u} = \nabla^2 \phi = 0$. Using Gauss's theorem, the first term gives:

$$E = \frac{1}{2} \int_S \phi \, (\mathbf{n} \cdot \nabla \phi) \, dS,$$

where $\mathbf{n}$ denotes a unit vector normal to the surface element dS.

2. Let $\mathbf{u}_1 = \nabla \phi_1$ and $\mathbf{u}_2 = \nabla \phi_2$ be two different solutions. Then $\mathbf{u} = \mathbf{u}_1 - \mathbf{u}_2$ is also a solution which satisfies $\mathbf{u} \cdot \mathbf{n} = 0$ at the boundary. Then, by part 1 $E = 0$ i.e. $\mathbf{u}(\mathbf{x}) \equiv 0$. Hence we arrive at a contradiction: $\mathbf{u}_1$ and $\mathbf{u}_2$ cannot be different solutions.

3. No, because the specified volume is not simply connected.

Chapter 3

Fluid with free surface

3.1 Background theory

Fluids often have a free surface, for example, an upper surface in systems under gravity, as commonly seen in everyday life when we pour water in a glass or when we look at the sea. As with solid surfaces, the role of the moving boundary will be in imposing boundary conditions on the fluid motion. However, in this case the boundary conditions will be different. Below we will consider the boundary conditions which are specific for the free-surface flows.

3.1.1 Pressure boundary condition

One condition is that the pressure of the fluid at the surface is the same as the pressure of the surrounding air—else the surface particles would experience infinite forces. The second condition is that the fluid particles, once at the surface, remain on the surface for indefinite time; it is called the kinematic boundary condition.

Let the gravity acceleration vector be in the negative z direction, $\mathbf{g} = g z$. Let the free fluid surface elevation be described by a function h marking the height at which the surface is $z = h(x, y, t)$. Then the pressure boundary condition will be

$$p(x, y, h(x, y, t)) = p_0, \tag{3.1}$$

where p_0 is the atmospheric pressure. For example, for irrotational flow one can use the time-dependent Bernoulli equation (2.5) and thus the pressure boundary condition becomes:

$$\left[\partial_t \phi + \frac{(\nabla \phi)^2}{2} \right]_{z=h} + gh = 0, \tag{3.2}$$

where we have chosen the constant $C = \frac{p_0}{\rho}$, which corresponds to a calibration condition that $\phi = 0$ in the motionless state.

Note that the pressure boundary condition is relevant to the cases where the fluid (or gas) beyond the free surface has much smaller density than the one of the main fluid and not-too-high velocity or then viscosity is unimportant. In both of these cases the pressure is the only important component of the stress tensor at the surface. If these conditions are not satisfied, the pressure boundary condition has to be replaced by a more general condition of continuity at the surface of all the stress tensor components.

3.1.2　Kinematic boundary condition

To obtain the kinematic boundary condition, let us introduce the function $f = h(x, y, t) - z$. The level set $f = 0$ marks the free surface. The statement that the surface particles remain on the surface can be formulated as conservation of function f along the fluid paths of the surface particles, i.e.

$$[\partial_t + (\mathbf{u} \cdot \nabla)]f(x, y, z, t) = 0, \tag{3.3}$$

or, after substituting for f, we finally get

$$\partial_t h - u_z + (\mathbf{u}_\perp \cdot \nabla)h = 0, \tag{3.4}$$

where $\mathbf{u}_\perp$ is the horizontal velocity projection, $\mathbf{u}_\perp = (u_x, u_y)$.

In the problems below we will only consider steady flows without a vertical velocity component. For these flows the kinematic boundary condition (3.4) is reduced to

$$(\mathbf{u}_\perp \cdot \nabla)h = 0, \tag{3.5}$$

i.e. the level sets of h coincide with the streamlines.

However, in the chapters devoted to waves and instabilities we will use the full unsteady version of the kinematic boundary condition (3.4).

It is worth mentioning that the boundary conditions on solid boundaries of the containing volume remain the same as for the cases without a free surface, namely they are the no-slip or the free-slip boundary conditions depending on whether we deal with a viscous or an inviscid fluid.

3.1.3　Axially and spherically symmetric flows

In all of the problems considered in the present chapter, there is either axial or spherical symmetry present. For axially symmetric 3D flows it is often useful to use the following expressions for the Laplacian and the divergence operators in cylindrical coordinates:

$$\Delta f = \frac{1}{r}\frac{\partial}{\partial r}\left(r\frac{\partial f}{\partial r}\right) + \frac{1}{r^2}\frac{\partial^2 f}{\partial \theta^2} + \frac{\partial^2 f}{\partial z^2},$$

$$\nabla \cdot \mathbf{u} = \frac{1}{r}\frac{\partial(r u_r)}{\partial r} + \frac{1}{r}\frac{\partial u_\theta}{\partial \theta} + \frac{\partial u_z}{\partial z}.$$

The radial component of $(\mathbf{u} \cdot \nabla)\mathbf{u}$ is

$$u_r \frac{\partial u_r}{\partial r} + \frac{u_\theta}{r}\frac{\partial u_r}{\partial \theta} + u_z \frac{\partial u_r}{\partial z} - \frac{u_\theta^2}{r}.$$

The spherically symmetric flows considered in this chapter will only have the radial velocity component. For considering these flows, it will be useful to use the equation for the radial component of velocity, v, which in this case is,

$$\partial_t v + v \partial_r v = -\frac{1}{\rho} \partial_r p. \qquad (3.6)$$

The incompressibility condition in this case is

$$\partial_r (r^2 v) = 0. \qquad (3.7)$$

3.2 Further reading

Discussions of the kinematic and the pressure boundary conditions at the free surface are frequently found in the context of the water wave theory and, therefore, we refer the reader to the specialised books about waves *Waves in Fluids* by J. Lighthill [15] and *Linear and Nonlinear Waves* by G.B. Whitham [31], as well as more general books *Elementary Fluid Dynamics* by D.J. Acheson [1], *Prandtl's Essentials of Fluid Mechanics* by Oertel et al [17] and *Fluid Dynamics* by L.D. Landau and E.M. Lifshitz [14].

3.3 Problems

3.3.1 Water surface distortion due to vortex

On a sunny day, vortices can be seen in a shallow creek as round dark spots at the creek's bottom. This is because vortices produce concave distortions on the water surface, which act to spread light rays apart. In this problem, you are asked to find the shape of the free surface of an ideal fluid in the presence of a *Rankine vortex*; see figure 3.1. Far from the vortex, the surface tends to its undisturbed shape, $h \to h_\infty$ for $r \to \infty$. You are reminded that a Rankine vortex has uniform vorticity $\omega = (0, 0, \Omega)$ ($\Omega =$const) inside a vertical cylinder of radius a (i.e. for $r = \sqrt{x^2 + y^2} < a$) and zero vorticity $\omega = (0, 0, 0)$ outside of this cylinder (for $r \geq a$).

Hint: You may find it useful to keep in mind that the uniformly rotating fluid also has constant vorticity distribution, so you should expect the Rankine vortex solution for $r < a$ to be the same as in a rotating bucket of water. Also remember that the surface must be continuous at $r = a$.

1. Derive the velocity field for this Rankine vortex for both $r < a$ and $r \geq a$.

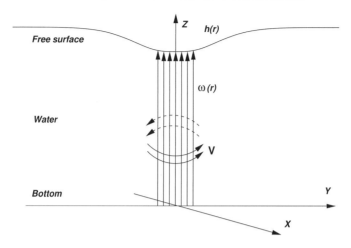

FIGURE 3.1: Vortex in a layer of fluid with free surface.

2. The pressure at the free surface is constant (atmospheric pressure). Can we use Bernoulli's theorem to find the shape of the water surface for $r < a$? For $r \geq a$? Explain why.

3. Find the shape of the water surface $z = h(r)$ for both $r < a$ and $r \geq a$ regions. Use Bernoulli's theorem wherever applicable, otherwise solve the ideal flow (Euler) equations directly. Sketch the water surface shape based on the mathematical expressions you have obtained.

3.3.2 Free surface shape of water in an annular flow

An incompressible viscous fluid fills an annular space between two vertical coaxial cylinders, whose inner and outer surfaces have radii a and R respectively; see figure 3.2. The system is subject to gravity and its upper surface is open to the atmosphere with pressure p_0. The inner cylinder is rotating at a constant angular velocity Ω and the outer cylinder is stationary. The fluid flow has reached a steady state in which its velocity is purely azimuthal and dependent on the radius r only, $u_r = u_z = 0$, $u_\theta = u_\theta(r)$. This problem deals with finding the shape of the free surface, namely its height profile $h(r)$.

1. Formulate the no-slip boundary conditions for this system. Formulate the pressure boundary condition at the free surface.

2. Show that the incompressibility condition is satisfied.

3. Find the steady-state velocity profile $u_\theta(r)$ that satisfies the Navier-Stokes equations with the no-slip boundary conditions at the moving cylinders.

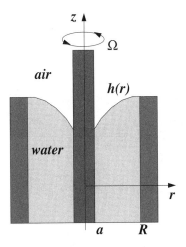

FIGURE 3.2: Water in an annular flow.

4. Find the pressure inside the fluid $p = p(r, z)$.

5. Use the pressure boundary condition and find the shape of the free surface $h(r)$.

3.3.3 Water-filled turntable

The Coriolis turntable in Grenoble is the world's largest rotating platform dedicated to fluid dynamics. It is a disk-like cylindrical tank of 13 m in diameter filled with water to about 1.2 m height. In this problem, the aim is to find the surfaces of constant pressure, and hence the shape of the free surface of a uniformly rotating cylinder of water which is interfacing at the top with air at atmospheric pressure; see figure 3.3.

An ideal fluid fills a cylindrical container to height h_0. The total height of the cylinder walls is $2h_0$ and its radius is R. The gravity field is in the negative z-direction. The system is rotating with a constant angle velocity Ω, so that in the laboratory frame the velocity field is $\mathbf{u} = (-\Omega y, \Omega x, 0)$.

1. One might argue that by Bernoulli's theorem, $p/\rho + u^2/2 + gz$ is constant, so the constant pressure surfaces are,

$$z = \text{constant} - \frac{\Omega^2}{2g}(x^2 + y^2).$$

But this means that the surface of a rotating water tank is at its highest in the middle, which is obviously not true. What is wrong with this argument?

2. Show that the velocity field $\mathbf{u} = (-\Omega y, \Omega x, 0)$ satisfies the incompressibility condition.

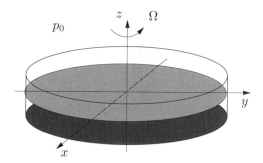

FIGURE 3.3: Rotating tank of water.

3. Write down the ideal fluid equations in component form, integrate them directly to find the pressure p and hence obtain the correct shape for the free surface.

4. Suppose the rotation speed is gradually increased. Which of the two events would happen first: (i) the water surface hits the bottom of the container or (ii) the water will spill over the edges of the container walls.

5. What would happen to the free surface if we introduce a viscosity ν?

3.3.4 Cavity implosion

Consider the motion of an ideal fluid which arises after a sudden removal of a spherical volume of fluid of radius a at time $t = 0$. Initially the fluid is at atmospheric pressure p_0, and the pressure inside the cavity is zero (vacuum). In this problem, we will study how the cavity shrinks by assuming that the velocity is always directed towards the cavity centre and depends only on the distance from the cavity centre; see figure 3.4. We will also neglect the role of gravity. (**Hint**: work with spherical coordinates—see the equations (3.6) and (3.7).)

1. Integrate the incompressibility condition with respect to the radius r (the result should involve an arbitrary function of time because no boundary condition is used at this point).

2. Substitute the results of part 1 into the equation for the radial velocity and integrate it over the radius r from the cavity radius $R(t)$ to infinity assuming that pressure remains atmospheric at $r = \infty$.

3. Consider the rate of change of the hole radius, $V = dR/dt$, and use results of parts (1) and (2) to show that function $V(R)$ satisfies the following differential equation,

$$\frac{3V^2}{2} + \frac{1}{2}R\frac{dV^2}{dR} = -\frac{p_0}{\rho}.$$

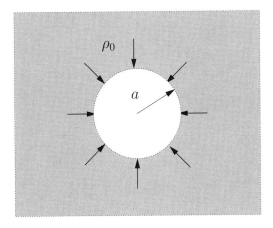

FIGURE 3.4: Cavity implosion.

4. Integrate the differential equation obtained in part (3) and thereby find an expression for V in terms of R. (**Hint**: use the boundary condition $V(a) = 0$ which means that the fluid is at rest initially.)

5. Find the time needed for the cavity to be filled with fluid. Leave your expression in terms of an integral over R (you do not have to evaluate this integral).

3.3.5 Flow in an expanding air bubble

Consider a flow produced by a radially expanding spherical air bubble in an infinite three-dimensional volume of incompressible fluid. The radius of the bubble r_0 is increasing with a given speed $V_0(t)$. The fluid motion is purely radial; see figure 3.5.

1. Find the fluid velocity field using the incompressibility condition.

2. Prove that the flow is purely irrotational.

3. The local rate $\epsilon(\mathbf{x}, t)$ at which the kinetic energy dissipates in the fluid is given by

$$2\rho\nu \sum_{i,j=1}^{3} (e_{i,j})^2,$$

where ρ and ν are the fluid density and the kinematic viscosity coefficient respectively, and

$$e_{i,j} = \frac{1}{2} \left(\frac{\partial u_i}{\partial x_j} + \frac{\partial u_j}{\partial x_i} \right)$$

is the rate of strain tensor.

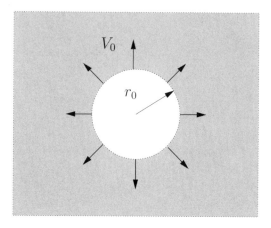

FIGURE 3.5: Expanding air bubble.

Find $\epsilon(\mathbf{x}, t)$ and, integrating it over the volume occupied by the fluid, find the total energy dissipated in the fluid per unit time.

3.3.6 Flow in wire coating die

A wire coating die consists of a cylindrical wire of radius a moving horizontally at a constant velocity V along the axis of a cylinder of radius R and length L. If the pressure in the die is uniform, then the viscous incompressible fluid flows through the narrow annular region solely by the drag due to the motion of the wire; see figure 3.6.

1. Formulate the no-slip boundary conditions on the die and on the wire.

2. Establish the expression for the steady-state velocity profile in the annular region of the die.

3. Obtain the mass flow rate through the annular die region.

4. Find the coating thickness b far downstream of the die's exit.

5. Find the force that must be applied to the wire.

3.4 Solutions

3.4.1 Model solution to question 3.3.1

1. From symmetry, the Rankine vortex velocity will be purely azimuthal. It can be found from considering the circulation integral over a circle of

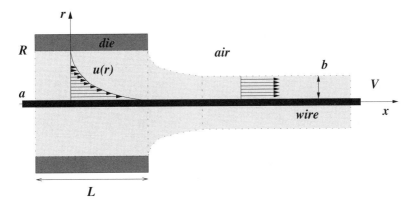

FIGURE 3.6: Flow in a wire coating die.

radius r centred at the vortex centre;

$$\oint_{C_r} u_\phi \, dl = \int_{D_r} \omega(r) \, d^2\mathbf{x},$$

where D_r is the disc bounded by circle C_r with radius r.

The left-hand side here is equal to $2\pi r u_\phi$ (because u_ϕ is constant on C_r), and the right-hand side is to be found using cylindrical coordinates. This gives:

$$u_\phi = \frac{\Omega r}{2} \quad \text{for} \quad r < a,$$

$$u_\phi = \frac{\Omega a^2}{2r} \quad \text{for} \quad r \geq a.$$

2. Bernoulli's theorem can be used for $r > a$ because this region is irrotational ($\omega = 0$). Region $r \leq a$ is not irrotational, and Bernoulli's theorem says that Bernoulli's potential is constant along streamlines, but it can have different value on different streamlines. This information is useless for finding the solution for the surface shape at $r \leq a$.

3. First we find a solution for $r > a$ from the Bernoulli's theorem at the fluid surface taking into account that the pressure at the surface $z = h(r)$ is constant equal to the atmospheric pressure p_0:

$$\frac{u^2}{2} + \frac{p}{\rho} + gh(r) = const.$$

Substituting u we have:

$$h(r) = h_\infty - \frac{\Omega^2 a^4}{8gr^2} \quad \text{for} \quad r \geq a,$$

where h_∞ is the height of undisturbed water surface.

To find solution for $r \leq a$, we write the steady-state Euler equation in components,

$$
\begin{aligned}
(\mathbf{u} \cdot \nabla)u &= -\partial_x p/\rho, \\
(\mathbf{u} \cdot \nabla)v &= -\partial_y p/\rho, \\
0 &= -\partial_z p/\rho - g;
\end{aligned}
$$

or

$$
\begin{aligned}
x\Omega^2 &= -\partial_x p/\rho, \\
y\Omega^2 &= -\partial_y p/\rho, \\
0 &= -\partial_z p/\rho - g.
\end{aligned}
$$

Integrating, we have

$$
\frac{p}{\rho} = \frac{\Omega^2}{2}(x^2 + y^2) - gz + const.
$$

On the surface we have

$$
h(r) = \frac{\Omega^2 r^2}{2g} + h_0 \quad \text{for} \quad r < a.
$$

By matching at $r = a$ we find $h_0 = h_\infty - \frac{5\Omega^2 a^2}{8g}$.

3.4.2 Model solution to question 3.3.2

1. No-slip boundary conditions: fluid velocity at the boundaries coincides with the one of the boundaries

$$
u_\theta(a) = \Omega a, \quad u_\theta(R) = 0.
$$

At the free surface

$$
p(r, h(r)) = p_0.
$$

2.

$$
\nabla \cdot \mathbf{u} = \frac{1}{r}\frac{\partial(ru_r)}{\partial r} + \frac{1}{r}\frac{\partial u_\theta}{\partial \theta} + \frac{\partial u_z}{\partial z}.
$$

Now, $u_r = u_z = 0$ and u_θ is independent of θ, so $\nabla \cdot \mathbf{u} = 0$.

3. Note that u_θ is constant along the streamlines in this flow, $D_t u_\theta = 0$. Pressure is independent of θ. Thus, the steady-state Navier-Stokes for the azimuthal component is

$$
0 = \nu \frac{1}{r}\partial_r (r\partial_r u_\theta).
$$

Solving this equation we have

$$u_\theta = \frac{C_1}{r} + C_2 r.$$

Matching to the boundary condition gives

$$u_\theta = \frac{\Omega a^2 R}{R^2 - a^2} \left(\frac{R}{r} - \frac{r}{R} \right).$$

4. The steady-state Navier-Stokes equation in the radial component is

$$\frac{u_\theta^2}{r} = \frac{1}{\rho} \partial_r p.$$

Substituting here the solution for u_θ and integrating, we get

$$p = \frac{\rho \Omega^2 a^4 R^2}{(R^2 - a^2)^2} \left(-\frac{R^2}{2r^2} - 2\ln r + \frac{r^2}{2R^2} \right) + f_1(z).$$

For the z-component:

$$0 = -\frac{1}{\rho} \partial_z p - g,$$

or $p = -\rho g z + f_2(r)$. Combining, we have

$$p = \frac{\rho \Omega^2 a^4 R^2}{(R^2 - a^2)^2} \left(-\frac{R^2}{2r^2} - 2\ln r + \frac{r^2}{2R^2} \right) - \rho g z + C_3.$$

5. Using the free-surface pressure condition,

$$h(r) = h(R) + \frac{\Omega^2 a^4 R^2}{g(R^2 - a^2)^2} \left(-\frac{R^2}{2r^2} - 2\ln \frac{r}{R} + \frac{r^2}{2R^2} \right).$$

3.4.3 Model solution to question 3.3.3

1. This version of the Bernoulli's theorem would only be valid for irrotational flows. Our flow is not irrotational because it has finite vorticity, $|\omega| = 2\Omega$.

2. We have $\nabla \cdot \mathbf{u} = -\Omega \partial_x y + \Omega \partial_y x = 0$.

3. The steady-state Euler equation in components:

$$\begin{aligned}
(\mathbf{u} \cdot \nabla)u &= -\partial_x p / \rho, \\
(\mathbf{u} \cdot \nabla)v &= -\partial_y p / \rho, \\
0 &= -\partial_z p / \rho - g;
\end{aligned}$$

or

$$
\begin{aligned}
x\Omega^2 &= -\partial_x p/\rho, \\
y\Omega^2 &= -\partial_y p/\rho, \\
0 &= -\partial_z p/\rho - g.
\end{aligned}
$$

Integrating, we have

$$
\frac{p}{\rho} = \frac{\Omega^2}{2}(x^2 + y^2) - gz + C,
$$

where $C = \text{const}$. On the surface we have

$$
h(r) = C_1 + \frac{\Omega^2 r^2}{2g}
$$

with $C_1 = C - \frac{p_0}{\rho}$. The fluid volume is conserved,

$$
V = 2\pi \int_0^R h(r)\, r\, dr = 2\pi \int_0^R \left[C + \frac{\Omega^2 r^2}{2g} \right] r\, dr =
$$

$$
\pi R^2 C_1 + \frac{\pi \Omega^2 R^4}{4g} = \text{const} = \pi R^2 h_0,
$$

so

$$
C_1 = h_0 - \frac{\Omega^2 R^2}{4g},
$$

and finally

$$
h(r) = h_0 + \frac{\Omega^2 (2r^2 - R^2)}{4g}.
$$

4. We have to compare $h(0)$ and $h(R)$:

$$
h(0) = h_0 - \frac{\Omega^2 R^2}{4g} \quad \text{and} \quad h(R) = h_0 + \frac{\Omega^2 R^2}{2g}.
$$

Thus, the surface drop in the centre is exactly twice less than the surface rise at the edges, which means that in our case the spill should occur before touching the bottom.

5. Viscosity would not change the results because the velocity field is linear in $\mathbf{x}$ and, therefore, its Laplacian is zero.

3.4.4   Model solution to question 3.3.4

1. From the incompressibility condition $\partial_r(r^2 v) = 0$. Integrating this equation we have $r^2 v = f(t)$, where f is an arbitrary function.

2. The radial component of the Euler equation is

$$\partial_t v + v \partial_r v = -\frac{1}{\rho} \partial_r p.$$

Substituting in here for v, we have:

$$\frac{f'}{r^2} + \frac{1}{2} \partial_r v^2 = -\partial_r (p/\rho).$$

Integrating over r from R to infinity, we have:

$$-\frac{f'}{R} + \frac{1}{2} V^2 = p_0/\rho.$$

We have used the boundary conditions $v(R) = V, v(\infty) = 0, p(R) = 0$ and $p(\infty) = p_0$.

3. For the function f we have $f = vr^2 = VR^2$ differentiating which we obtain:

$$f' = V'(R)R^2 R' + V2RR' = V'VR^2 + 2V^2 R.$$

Substituting this into part 2, we have:

$$-\frac{3}{2} V^2 - \frac{V'R^2}{R} V = p_0/\rho$$

or

$$3V^2 + R\frac{dV^2}{dR} = -2p_0/\rho.$$

4. Separating variables V^2 and R,

$$\frac{dV^2}{\frac{2p_0}{3\rho} + V^2} = -3\frac{dR}{R},$$

and integrating, we have:

$$\frac{2p_0}{3\rho} + V^2 = \frac{C}{R^3}.$$

where constant C is to be found from $V(a) = 0$. This gives $C = \frac{2p_0 a^3}{3\rho}$, so that

$$V^2 = \frac{2p_0}{3\rho}(a^3/R^3 - 1).$$

We choose $V < 0$, therefore

$$V = -\sqrt{\frac{2p_0}{3\rho}(a^3/R^3 - 1)}.$$

5. Integrating, we find for the implosion time:

$$\tau = \int_0^a \frac{dR}{V} = \sqrt{\frac{2p_0}{3\rho}} \int_0^a \frac{dR}{\sqrt{a^3/R^3 - 1}}.$$

3.4.5 Model solution to question 3.3.5

1. Because of the incompressibility condition, the volume flux through the spheres of different radii is the same. Hence:

$$u_r = V_0 r_0^2 / r^2.$$

2. In Cartesian frame

$$\mathbf{u} = (u, v, w) = \left(\frac{V_0 r_0^2 x}{r^3}, \frac{V_0 r_0^2 y}{r^3}, \frac{V_0 r_0^2 z}{r^3} \right).$$

The flow is purely irrotational:

$$\omega_z = (\nabla \times \mathbf{u})_z = \partial_x v - \partial_y y = -\frac{3V_0 r_0^2 x 2y}{2r^5} + \frac{3V_0 r_0^2 y 2x}{2r^5} = 0$$

and the same for ω_x and ω_y.

3. Since $\nabla \times \mathbf{u} = 0$, we have $\partial_i u_j = \partial_j u_i$. Therefore

$$e_{ij} = \frac{1}{2} \left(\frac{\partial u_i}{\partial x_j} + \frac{\partial u_j}{\partial x_i} \right) = \frac{\partial u_i}{\partial x_j} = \frac{V_0 r_0^2}{r^3} \left(\delta_{ij} - 3\frac{x_i x_j}{r^2} \right).$$

Thus, for the kinetic energy density dissipation rate we have:

$$\epsilon(\mathbf{x}, t) = 2\rho\nu \sum_{i,j=1}^{3} (e_{i,j})^2 = 2\rho\nu \sum_{i,j=1}^{3} \frac{V_0^2 r_0^4}{r^6} \left(\delta_{ij}\delta_{ij} - 6\delta_{ij}\frac{x_i x_j}{r^2} + 9\frac{x_i^2 x_j^2}{r^4} \right) =$$

$$12\rho\nu \frac{V_0^2 r_0^4}{r^6}.$$

Integrating it over the volume occupied by the fluid, we find the total energy dissipated in the fluid per unit time:

$$\dot{E} = 4\pi \int_{r_0}^{\infty} r^2 \epsilon(\mathbf{x}, t) \, dr = 16\pi\rho\nu V_0^2 r_0.$$

3.4.6 Model solution to question 3.3.6

1. The no-slip boundary conditions are $u_x(R) = 0, \ u_x(a) = V.$

2. The flow is stationary and it is not pressure driven, $\nabla p = 0$. Then the x-component of Navier-Stokes equation becomes:

$$\nu \frac{1}{r} \partial_r (r \partial_r u_x) = 0.$$

Integrating twice we find:

$$u_x = C_1 \ln r + C_2.$$

Using the no-slip boundary conditions we have:

$$u_x = V \frac{\ln \frac{r}{R}}{\ln \frac{a}{R}}.$$

3. We have for the mass flux:

$$Q = \rho \int_a^R u_x 2\pi r dr = \frac{2\pi \rho V}{\ln \frac{a}{R}} \int_a^R r \ln \frac{r}{R} dr =$$

$$\frac{\pi \rho V R^2}{2 \ln \frac{R}{a}} \left[1 + \frac{a^2}{R^2} \left(2 \ln \frac{a}{R} - 1 \right) \right].$$

4. $Q = \rho V \pi (b^2 - a^2)$, so $b = \sqrt{a^2 - Q/(\pi \rho V)}$.

5. Force = stress × area, so

$$F = \rho \nu |\partial_r u_x|_{r=a} (2\pi a L) = \frac{2\pi \rho \nu V L}{\ln \frac{R}{a}}.$$

Chapter 4

Waves and instabilities

4.1 Background theory

4.1.1 Waves

Waves represent one of the most common types of fluid motion: sea waves on water surface, internal waves in stratified oceans and atmospheres (often seen in cloud patterns), inertial waves in rotating fluids, Rossby waves of planetary scales important for the weather and climate dynamics, sound waves in air and many more important examples.

The basic element of the wave dynamics in most applications is the simplest type of wave, *monochromatic wave*: it is a fluid motion in which all relevant physical fields $a(\mathbf{x}, t)$ behave like

$$a(\mathbf{x}, t) = A \cos\left(\mathbf{k} \cdot \mathbf{x} - \omega t + \varphi\right), \qquad (4.1)$$

where $A \in \mathbb{R}_+$ is the wave amplitude, $\mathbf{k} \in \mathbb{R}^d$ is the wave vector in a d-dimensional space (e.g. $d = 2$ for the water surface waves, $d = 3$ for the sound waves in air, etc.), $\varphi \in \mathbb{R}$ is the wave phase and $\omega \in \mathbb{R}$ is the angular wave frequency (measured in radians per second), which thereafter will be called simply the frequency. The frequency is related to $\mathbf{k}$ via a *dispersion relation*,

$$\omega = \omega(\mathbf{k}), \qquad (4.2)$$

the form of which depends on the particular type of waves. For isotropic systems ω depends only on the magnitude of the wave vector, $k = |\mathbf{k}|$, and not on its direction. If for isotropic systems ω is a linear function of $k = |\mathbf{k}|$ (like e.g. for sound waves), the waves are called *non-dispersive*. For anisotropic systems, the waves are called non-dispersive if ω is a linear function of k times an arbitrary function of the unit vector along $\mathbf{k}$ (orientation). Otherwise the waves are called dispersive. The origin of this name will become clear when we solve question 4.3.1.

Note that different fluid quantities, such as the pressure, density and different velocity components, will have their own interrelated amplitudes A and phases φ. This is where one finds a possibility of various wave *polarisations*—plane, circular, elliptic—which refers to the shape of the curve traced in time by the tip of the velocity vector at a fixed point $\mathbf{x}$ in the course of one wave oscillation. On the other hand, frequency ω is the same for all different fluid quantities (for a fixed value of $\mathbf{k}$).

Wavelength λ is the distance between the adjacent wave crests; it is related to the wave vector by $\lambda = 2\pi/k$.

Representing the cosine function in terms of the complex exponentials, we can write (4.1) as

$$a(\mathbf{x}, t) = B\, e^{i(\mathbf{k} \cdot \mathbf{x} - \omega t)} + c.c., \qquad (4.3)$$

where *c.c.* means complex conjugate, amplitude B is now complex and contains both the wave magnitude and the phase,

$$B = \frac{1}{2} A\, e^{i\varphi}.$$

A monochromatic wave is, of course, an idealisation because it is infinite in the physical space. A slightly more realistic object is a *wave packet* which locally looks like the monochromatic wave but the amplitude of which decays with the distance from its centre; see figure 4.1. We will analyse wave packets in question 4.3.1 and establish that the wave packet centre moves with so-called *group velocity* given by

$$\mathbf{c}_g = \frac{\partial \omega(\mathbf{k})}{\partial \mathbf{k}}, \qquad (4.4)$$

whereas its individual wave crests move with so-called *phase velocity*,

$$\mathbf{c}_{ph} = \frac{\omega(\mathbf{k})}{|\mathbf{k}|} \frac{\mathbf{k}}{|\mathbf{k}|}. \qquad (4.5)$$

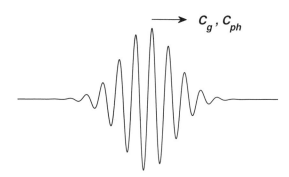

FIGURE 4.1: Wave packet.

The wave packet's wave vector $\mathbf{k}$ may also change in time as it moves through physical space. This happens if the frequency ω explicitly depends on the physical coordinate $\mathbf{x}$, i.e. $\omega = \omega(\mathbf{k}, \mathbf{x})$, which means that the wave packet is moving in a non-uniform background, e.g. a medium with inhomogeneous density and/or velocity field.

In physical space, one can represent the wave packet as:

$$a(\mathbf{x}, t) = \mathcal{A}(\epsilon \mathbf{x}, \epsilon t) \, e^{\frac{i}{\epsilon} \psi(\epsilon \mathbf{x}, \epsilon t)} + c.c., \tag{4.6}$$

where $\mathcal{A} \in C$ is a slowly varying wave amplitude and $\frac{1}{\epsilon}\psi \in R$ is the wave phase. The formal small parameter ϵ is introduced here to make the space-time dependence in $\mathcal{A}$ slow and to make the phase values large (so that the phase derivatives are order one). Without loss of generality one can take $\psi(0,0) = 0$. Indeed, to remove a non-zero value of $\psi(0,0)$ one could simply redefine $\mathcal{A}$ so that in the leading (ϵ^0) order the phase is a linear function of $\mathbf{x}$ and t. Then, in the leading order we recover the monochromatic wave (4.3) and the dispersion relation (4.2), in which

$$\mathbf{k} = \frac{1}{\epsilon}\nabla\psi(\epsilon\mathbf{x}, \epsilon t) \quad \text{and} \quad \omega = \frac{1}{\epsilon}\partial_t\psi(\epsilon\mathbf{x}, \epsilon t). \tag{4.7}$$

But relations (4.7) imply that for the wave packet

$$\partial_t \mathbf{k} = -\nabla\omega. \tag{4.8}$$

In the next (ϵ^1) order one gets the so-called transport equation governing the evolution of the amplitude $\mathcal{A}$, but we will not consider it here.

Changes in $\mathbf{k}$ would in turn lead to changes in the group velocity $\mathbf{c}_g$: the wave packet may slow down or accelerate or change direction of its propagation up to turning back (reflection). Finding the wave packet trajectory in the $(\mathbf{x}, \mathbf{k})$-space is called ray tracing. It is done by solving the *ray tracing equations* which follow from equations (4.5) and (4.8):

$$\dot{\mathbf{x}}(t) = \frac{\partial\omega(\mathbf{k}, \mathbf{x})}{\partial\mathbf{k}}, \tag{4.9}$$

$$\dot{\mathbf{k}}(t) = -\nabla\omega(\mathbf{k}, \mathbf{x}), \tag{4.10}$$

where the dot means the time derivative. Note that here $\mathbf{x}(t)$ and $\mathbf{k}(t)$ are time-dependent vectors and not an independent coordinate $\mathbf{x}$ and a vector field $\mathbf{k}(\mathbf{x}, t)$ as before.

4.1.2 Instabilities

The concept of stability of an equilibrium state is very natural in our everyday life, and it can be most easily illustrated by an example given in figure 4.2. On the left we see a ball at the bottom of a dimple. If we push the ball slightly out of equilibrium, it will experience a restoring force acting toward the equilibrium point: this equilibrium is stable. On the right we see a ball on the top of a hill. Slight disturbance will set this ball in a motion that will take it away from the equilibrium point: this equilibrium is unstable.

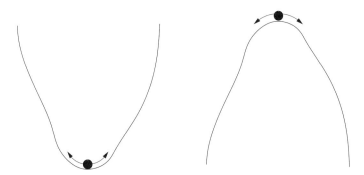

FIGURE 4.2: Stable (left) and unstable (right) fixed points.

Generalising this example, we say that a fixed (equilibrium) point is stable if for an arbitrarily chosen vicinity of this fixed point, all trajectories originating close enough to the fixed point will remain within the chosen vicinity. If the trajectory is converging to the fixed point, then the stability is asymptotic. If the solution is diverging from the fixed point (the distance to it is increasing in time), then this equilibrium is unstable. Note that in our example on the left of figure 4.2 the stability is asymptotic if friction is present: in this case the ball will be converging to the bottom point in time. Without friction, the ball will keep oscillating with non-decaying amplitude, so the stability is not asymptotic. Also note that in our example on the right of figure 4.2 the instability occurs for any initial disturbance, no matter how small. This type of instability is usually studied using a linearisation of the dynamical equations with respect to the small disturbances, and therefore it is called a *linear instability*.

One could slightly modify our example and consider a small dimple on the top of the hill, so that there would be a minimal disturbance necessary for setting the ball into the unstable motion. This kind of instability is called a finite-amplitude instability, and it is very important in the fluid dynamics context, but it is harder to study and we will omit it from the present book.

Instabilities play a central role in the fluid dynamics theory because of their widespread occurrences in nature, laboratory and industry. Fluid instabilities are usual precursors to turbulence—i.e. random fluid flows which are more common than laminar flows. An important difference of the fluid instability with the example above is that the equilibrium state is represented by a particular flow configuration, i.e. by continuous fields (velocity, density, pressure) rather than by finite dimensional coordinates.

Respectively, the perturbation of the equilibrium state is also a field, which in general makes the fluid instabilities harder to study than the instabilities in finite-dimensional systems, even in the linearised systems. This is because the linear equations, in general, may have coordinate-dependent coefficients. and their analytical solutions may be unavailable. Even worse, the respective linear operator may not be self-adjoint, which means that even if all the eigenmodes are stable, there could be algebraically growing disturbances.

Simplifications arise when the equilibrium flow is uniform (coordinate-independent) in one or several directions in physical space. Then an arbitrary linear disturbance can be represented as a linear combination of independent waves in this (these) direction(s). This approach is very similar to the one we employed in section 4.1.1 when we studied waves.

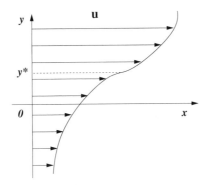

FIGURE 4.3: Plane-parallel shear flow.

Let us consider a classical setup for instability studies—plane-parallel flow with velocity $\mathbf{u} = (u(y), 0)$; see figure 4.3. This flow is a stationary solution for an ideal flow described by the inviscid flow (Euler) equations for any profile $u(y)$, but some profiles are known to be unstable. It is not possible to establish stability or instability for an arbitrary profile $u(y)$, but there exists a general necessary condition for an instability. Namely, the famous Rayleigh criterion (derivation of which can be found in many basic fluid dynamics texts) says that for an ideal plane-parallel flow between two plates (with free-slip boundary conditions) to be linearly unstable, the velocity profile $u(y)$ must have an inflection point—like the point marked by a dashed line in figure 4.3. However, this criterion only refers to exponentially growing modes associated with the eigenvalue problem, and it does not rule out a possibility of algebraically growing modes in flows without inflection points. In question 4.3.12 we will consider one such example: algebraic growth of disturbances of a constant-shear flow $u(y) = sy$, $s = \text{const}$ (there are no eigenmodes at all for this flow). We will also consider the most important example of a plane-parallel flow for which the full analytical solution of the eigenvalue problem is possible: a flow with constant velocities on both sides of an interface—the so-called Kelvin-Helmholtz instability; see problem 4.3.10. We will also consider an instability that arises when a heavy fluid is placed on top of a lighter fluid—the so-called Rayleigh-Taylor instability; see problem 4.3.11.

4.2 Further reading

Discussions of various aspects of waves and instabilities in fluids can be found in most Fluid Dynamics textbooks, e.g. in the books *An Introduction to Fluid Dynamics* by G.K. Batchelor [4], *Elementary Fluid Dynamics* by D.J. Acheson [1], *Prandtl's Essentials of Fluid Mechanics* by Oertel et al. [17] and *Fluid Dynamics* by L.D. Landau and E.M. Lifshitz [14].

More advanced discussions of the waves can be found in the classical books *Waves in Fluids* by J. Lighthill [15] and *Linear and Nonlinear Waves* by G.B. Whitham [31]. An excellent text on the fluid instabilities is the book *Hydrodynamic Stability* by P.G. Drazin and W.H. Reid [5]. The Rapid Distortion Theory of the question 4.3.12 is presented in detail in the book *The Structure of Turbulent Shear Flow* by A.A. Townsend [27].

4.3 Problems

4.3.1 Motion of a wave packet

Consider a *wave packet* which looks like the monochromatic wave whose amplitude is enveloped by a slow function which decays with the distance from its centre; see figure 4.1. Such a shape can be represented as a linear combination of monochromatic waves using the Fourier transform,

$$a(\mathbf{x}, t) = \int_{\mathbf{k} \in \mathbb{R}^d} \hat{b}(\mathbf{k})\, e^{i(\mathbf{k} \cdot \mathbf{x} - \omega(\mathbf{k})\, t)}\, d\mathbf{k} + c.c., \qquad (4.11)$$

where function $b(\mathbf{k})$ is narrowly peaked around some wave vector $\mathbf{k}_c$; see figure 4.4. We will refer to $\mathbf{k}_c$ as the carrier-wave vector.

1. Consider first an extreme case of a delta-function peak, $\hat{b}(\mathbf{k}) = B\, \delta(\mathbf{k} - \mathbf{k}_c)$, and recover the monochromatic wave expression (4.3).

2. Now take a slightly wider peak. If the peak is narrow enough, one could Taylor expand function $\omega(\mathbf{k})$ around $\mathbf{k}_c$. Find the condition on the characteristic width of function $\hat{b}(\mathbf{k})$ for such a Taylor expansion to be valid.

3. Perform the Taylor expansion of $\omega(\mathbf{k})$ around $\mathbf{k}_c$ up to the linear in $(\mathbf{k} - \mathbf{k}_c)$ term, and substitute it into the integral of expression (4.11). Represent the result as a product of the monochromatic carrier wave part $a_c = e^{i(\mathbf{k}_c \cdot \mathbf{x} - \omega(\mathbf{k}_c)\, t)}$ and an envelope part a_e. Find a_e and show that it varies in physical space much slower than the carrier wave part a_c.

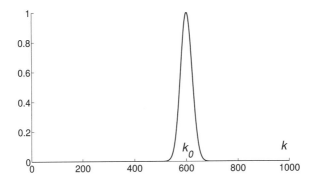

FIGURE 4.4: Fourier distribution in a wave packet.

4. By considering the $\mathbf{x}$ and t dependence of a_e, prove that the envelope (and therefore the wave packet as a whole) moves with the group velocity $\mathbf{c}_g = \frac{\partial \omega(\mathbf{k})}{\partial \mathbf{k}}$.

5. By considering the $\mathbf{x}$ and t dependence of a_c, prove that the wave crests move with the phase velocity $\mathbf{c}_{ph} = \frac{\omega(\mathbf{k})}{|\mathbf{k}|} \frac{\mathbf{k}}{|\mathbf{k}|}$.

6. Consider now for simplicity a one-dimensional wave packet, i.e. such that $\hat{b}(\mathbf{k}) \neq 0$ only for $\mathbf{k}$ that are in the same direction as $\mathbf{k}_c$. Consider the Taylor expansion of $\omega(\mathbf{k})$ around $\mathbf{k}_c$ beyond the linear term, establish which terms in the Taylor expansion lead to gradual evolution (dispersion) of the envelope shape a_e. Thus establish the condition for the waves to be non-dispersive, i.e. such that their envelope never changes its shape.

4.3.2 Gravity waves on the water surface

We will start with the most familiar example: gravity waves on the water surface.

Given information: kinematic boundary condition on the free surface (3.4).

1. Consider infinitely deep water in the gravity field (directed in the negative z direction) with free surface at $z = h(x, t)$ above which there is air the atmospheric pressure. Assume that the flow below the surface is irrotational and use the time-dependent version of Bernoulli's theorem to formulate the pressure boundary condition as an equation for the velocity potential ϕ at the water surface.

2. Consider a plane wave solution

$$h(x,t) = \frac{H}{2} e^{i(kx-\omega t)} + c.c., \quad \phi(x,z,t) = B(z) e^{i(kx-\omega t)} + c.c., \quad (4.12)$$

where H is a real constant (wave amplitude) and $B(z)$ is a complex function.

From the incompressibility condition and the condition that the velocity field must vanish as $z \to -\infty$, find the shape of the function $B(z)$.

3. Substitute (4.12) into the equations expressing the pressure and the kinematic boundary conditions and, assuming that the wave is weak, linearise these equations. Find the expressions of the frequency ω as a function of k and g, and the function $B(z)$ in terms of H, k and g (g being the gravity acceleration constant).

4. Find a condition on the wave amplitude H under which the linearisation procedure is justified. What does it mean in terms of the water surface inclination?

5. Find the phase and the group velocities of the gravity wave. What is the ratio of these velocities?

6. Find the shape of the fluid paths under the water surface.

4.3.3 Gravity and capillary waves: dimensional analysis

In this problem we will study waves on the water surface: gravity waves which appear at larger scales and capillary waves at smaller scales. We will not solve the fluid equations with free surface in presence of gravity and surface tension, but we will take a short-cut by postulating that the wave properties in the respective situations may only be determined by relevant dimensional physical constants: the gravity acceleration in the case of the gravity waves and the surface tension coefficient and the water density in the case of the capillary waves.

1. Consider water in the gravity field with free surface above which there is air the atmospheric pressure. Argue why the pressure of air above the free surface cannot enter the force balance of the moving fluid particles and, therefore, they will not enter into the characteristics of the water waves.

2. Consider a monochromatic wave on the free surface of infinitely deep water in presence of gravity (ignore the surface tension for now). Since the air pressure is irrelevant, and because there is no characteristic length scale in the infinite depth ocean at rest, the only other physical quantities that could determine the properties of the linear waves are the gravity constant g, the wave vector magnitude k and the water density ρ.

Find the physical dimensions of g, k and the wave frequency ω. Find the only possible combination of g, ρ and k which would have the dimension of ω and, therefore, find the dispersion relation $\omega = \omega(k)$ for the gravity waves. (As usual, such a dimensional method leaves uncertainty up to an order of one dimensionless constant, which in our case appears to be absent).

Does this relation depend on ρ? Why?

3. Now consider a monochromatic capillary wave on the free surface of infinitely deep water whose oscillations are due to the surface tension forces (ignore gravity). Find the physical dimension of the surface tension coefficient σ, based on its definition as a force per unit length (think of a force produced by a soap film onto a wire to which it is attached).

4. Find the unique combination of σ, ρ and k which would have the dimension of ω and, therefore, find the dispersion relation $\omega = \omega(k)$ for the capillary waves. (Again, the undetermined dimensionless constant, in this case appears to be absent).

5. Now consider a system in which both gravity and surface tension are present. Find the crossover wave length at which the capillary effect becomes more important than the gravity effect. Surface tension coefficient at room temperature is about 0.072 Newton per meter, and the gravity constant is $9.8\ m/s^2$. Use these values to find the crossover wave length.

4.3.4 Inertial waves in rotating fluids

> Given information: Euler equation in a uniformly rotating frame of reference (1.12) (with $\nu = \mathbf{f} = 0$).

This problem concerns the dynamics of inertial waves, i.e. small perturbations to the state of uniform rotation.

1. Write down an expression for the velocity field corresponding to uniform rotation. Find the vorticity corresponding to this flow.

2. Consider a small perturbation $\mathbf{u}$ to the state of uniform rotation (with an angular velocity $\mathbf{\Omega}$) which has the form of a harmonic wave

$$\mathbf{u} = \mathbf{A}\, e^{i(\mathbf{k}\cdot\mathbf{x} - \omega t)} + \mathbf{A}^*\, e^{-i(\mathbf{k}\cdot\mathbf{x} - \omega t)},$$

where $\mathbf{k}$ is the wave vector, ω is the frequency, $\mathbf{A}$ is the complex amplitude and $\mathbf{A}^*$ is the complex conjugate of $\mathbf{A}$. How small does the amplitude $\mathbf{A}$ need to be for the nonlinear (with respect to $\mathbf{A}$) terms to be much smaller than the linear ones in the fluid equations?

3. Consider the linearised equations of motion and derive the dispersion relation $\omega = \omega(\mathbf{k})$. (**Hint**: Take the curl of the Euler equation (1.12) and use the identity $\nabla \times (\boldsymbol{\Omega} \times \mathbf{u}) = -(\boldsymbol{\Omega} \cdot \nabla)\mathbf{u}$.)

4. What is the polarisation of the inertial waves?

5. Are these waves dispersive or non-dispersive? Isotropic or anisotropic? Explain why.

   Find the group velocity for the inertial waves and comment on its relative direction with respect to the wave number.

4.3.5 Internal waves in stratified fluids

In this problem we will study waves that arise due to stratification of the fluid density: the so-called internal waves. They play important roles in both atmospheric and ocean dynamics. In particular, in the oceans they provide the principal mechanism for bringing the global conveyor belt currents from the ocean's bottom to the surface, and as such they are one of the key ingredients in climate dynamics.

Equations for inviscid incompressible fluid in the presence of a density stratification are (1.5), (1.3) and (1.4). In the presence of gravity $\mathbf{f} = -g\,\hat{\mathbf{z}}$, these equations become

$$\partial_t \mathbf{u} + (\mathbf{u} \cdot \nabla)\mathbf{u} = -\frac{1}{\rho}\nabla p - g\,\hat{\mathbf{z}}, \tag{4.13}$$

$$\nabla \cdot \mathbf{u} = 0 \tag{4.14}$$

and

$$\partial_t \rho + \mathbf{u} \cdot \nabla \rho = 0. \tag{4.15}$$

Consider an incompressible fluid subject to gravity, which at rest has an exponential density profile,

$$\rho = \rho_0(z) = \rho_0(0)\, e^{-z/h}.$$

1. Find the equilibrium pressure profile $p_0(z)$.

2. Consider perturbations around the equilibrium state,

$$\rho = \rho_0(z) + \tilde{\rho}, \qquad \mathbf{u} = \tilde{\mathbf{u}},$$

   and linearise the fluid equations (4.13) and (4.15).

   Consider perturbations in the form of a harmonic wave

$$\tilde{\mathbf{u}} = \frac{\mathbf{A}}{\sqrt{\rho_0(z)}}\, e^{i(\mathbf{k}\cdot\mathbf{x}-\omega t)} + c.c., \tag{4.16}$$

$$\tilde{\rho} = R\sqrt{\rho_0(z)}\, e^{i(\mathbf{k}\cdot\mathbf{x}-\omega t)} + c.c., \tag{4.17}$$

$$\tilde{p} = P\sqrt{\rho_0(z)}\, e^{i(\mathbf{k}\cdot\mathbf{x}-\omega t)} + c.c., \tag{4.18}$$

where $\mathbf{k} \in \mathbb{R}^3$ is the wave vector, $\omega \in \mathbb{R}$ is the frequency, $\mathbf{A} \in \mathbb{C}^3$ is a constant vector and $R \in \mathbb{C}$ and $P \in \mathbb{C}$ are constant scalars; c.c. stands for "complex conjugate".

Substituting (4.16), (4.17) and (4.18) into the linearised equations, find the wave dispersion relation, $\omega = \omega(\mathbf{k})$.

3. Are these waves dispersive or non-dispersive? Isotropic or anisotropic? Explain why.

   Consider the short-wave limit $kh \gg 1$. Find the group velocity for the inertial waves and comment on its relative direction with respect to the wave vector.

4.3.6 Sound waves in compressible fluids

Equations for inviscid compressible flow are (1.5), (1.6) and (1.7). In absence of forcing ($\mathbf{f} = 0$) for isentropic motions ($S = \text{const}$) of polytropic gas, these equations become

$$\partial_t \mathbf{u} + (\mathbf{u} \cdot \nabla)\mathbf{u} = -\frac{1}{\rho}\nabla p, \tag{4.19}$$

$$\partial_t \rho + \nabla \cdot (\rho\mathbf{u}) = 0 \tag{4.20}$$

with

$$\frac{p}{\rho^\gamma} = \text{const}, \tag{4.21}$$

where $\gamma = \text{const}$ is the adiabatic index.

Consider a uniform gas at rest with $\rho = \rho_0 = \text{const}$, $p = p_0 = \text{const}$ and $\mathbf{u} = 0$. Small perturbations of such a state ($\tilde{\rho}$, $\tilde{p}$ and $\tilde{\mathbf{u}}$ respectively) propagate as acoustic waves (sound) which will be studied in the present problem.

1. Substitute the disturbed fields $\rho = \rho_0 + \tilde{\rho}$, $p = p_0 + \tilde{p}$ and $\mathbf{u} = \tilde{\mathbf{u}}$ into the equations (4.19), (4.20) and (4.21) and, assuming the perturbations to be small, linearise these equations.

2. Consider perturbations of the form of a harmonic wave

$$\tilde{\mathbf{u}} = \mathbf{A}\,e^{i(\mathbf{k}\cdot\mathbf{x}-\omega t)} + \mathbf{A}^*\,e^{-i(\mathbf{k}\cdot\mathbf{x}-\omega t)}, \tag{4.22}$$

$$\tilde{\rho} = B\,e^{i(\mathbf{k}\cdot\mathbf{x}-\omega t)} + B^*\,e^{-i(\mathbf{k}\cdot\mathbf{x}-\omega t)}, \tag{4.23}$$

   where $\mathbf{k}$ is the wave vector, ω is the frequency, $\mathbf{A}$ and B are the complex amplitude vector and scalar respectively, and $\mathbf{A}^*$ and B^* are their complex conjugates.

   Substituting (4.22) and (4.23) into the linearised equations, find the sound-wave dispersion relation, $\omega = \omega(\mathbf{k})$, as well as the relation between $\mathbf{A}$ and B.

3. How small does the amplitude $\mathbf{A}$ need to be for the linearisation of the fluid equations to be valid?

4.3.7 Sound rays in shear flows: acoustic mirage

Acoustic rays may be bent and even reflected in strong wind conditions, which may have an effect of sound amplification on the downwind side of the source and appearance of a quiet zone on the upwind side. This effect is similar to the mirage phenomenon arising due to deflection of light rays due to refraction index variations in an inhomogeneously heated air. In this problem we will study this effect using the ray tracing equations (4.9) and (4.10).

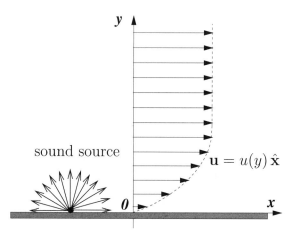

FIGURE 4.5: Sound propagation in a shear flow.

Consider a point sound source on the ground emitting acoustic wave packets in all upward facing directions; see figure 4.5. A strong wind above the ground has a form of a boundary layer with plane-parallel velocity profile $\mathbf{u} = (u(y), 0, 0)$ with

$$u(y) = \frac{2u_0}{\pi} \arctan(\delta y),$$

where y is the distance from the ground and δ and u_0 are positive constants— boundary layer thickness and velocity at high altitude respectively.

Sound frequency is modified in the presence of the wind by the Doppler effect:

$$\omega = c_s k + \mathbf{k} \cdot \mathbf{u},$$

where $k = |\mathbf{k}|$ and c_s is the speed of sound which we will assume to be constant in this problem.

1. Prove that the x-component of the wave number stays constant as the acoustic packet moves.

2. Prove that the frequency stays constant as the acoustic packet moves.

3. Find the y-component of the wave number as a function of the height y.

4. Find the condition for wave packet's reflection, i.e. for turning its trajectory back towards the ground. Find the height of the reflection layer $y = y_r$.

5. Sketch the acoustic rays. Explain the amplification of sound on the downwind side of the source and appearance of a quiet zone upwind of the source.

4.3.8 Sound rays in stratified flows: wave guides, Snell's law

When sound rays propagate in a medium with inhomogeneous pressure or/and temperature they deviate from straight lines and may reflect off layers with higher speed of sound. A quote from Wikipedia:

"The SOFAR channel (short for Sound Fixing and Ranging channel), or deep sound channel (DSC), is a horizontal layer of water in the ocean at which depth the speed of sound is at its minimum. The SOFAR channel acts as a waveguide for sound, and low frequency sound waves within the channel may travel thousands of miles before dissipating. This phenomenon is an important factor in submarine warfare."

This effect is similar to the refraction and reflection of light rays at the interfaces of changing refraction index. The relation between the angles of incidence and refraction is known in optics as Snell's law. Consequently, a similar Snell's law can be found for the angles of incidence and refraction of the acoustic rays, which will be the subject of the present problem.

Consider a stratified medium with the speed of sound being a monotonously increasing function of one space coordinate, e.g. height y: $c_s \equiv c_s(y)$. A sound wave packet is emitted at $y = 0$ with a positive y-component of the wave number.

1. Prove that the wave number component perpendicular to the y-axis stays constant as the acoustic packet moves.

2. Prove that the frequency stays constant as the acoustic packet moves.

3. Find the y-component of the wave number as a function of the height y.

4. At which position $y = y_r$ will the wave packet reflection occur?

5. Sketch the acoustic rays. Also sketch rays in a system which has a speed of sound minimum around some layer y_m and discuss this system in terms of its wave guiding properties.

6. Suppose now that the speed of sound changes abruptly at a sharp interface: $c_s(y) = c_1$ for $y < y_0$ and $c_s(y) = c_2$ for $y > y_0$, where y_0, c_1 and c_2 are positive constants and $c_2 > c_1$.

   Find the condition of reflection for a sound ray approaching the interface from $y < y_0$, as well as the relation between the angles of incidence and refraction (Snell's law).

4.3.9　Sound rays in a vortex flow: a black hole effect

Consider an acoustic wave packet propagating on background of a velocity field $\mathbf{u}(\mathbf{x})$ produced by a point vortex with circulation Γ:

$$\mathbf{u}(\mathbf{x}) = \frac{\Gamma}{2\pi r}\,\hat{\boldsymbol{\theta}},$$

where r is the distance from the vortex and $\hat{\boldsymbol{\theta}}$ is a unit vector in the azimuthal direction.

Interaction with the vortex field will bend the acoustic rays, which may be interpreted as a scattering process. Remarkably, under certain conditions the rays spiral in and fall onto the vortex, which is a fluid dynamics analogy for the black hole system. Below we will study this effect using the ray tracing equations (4.9) and (4.10).

Sound frequency is modified in presence of a velocity field by the Doppler shift:

$$\omega = c_s k + \mathbf{k} \cdot \mathbf{u},$$

where $k = |\mathbf{k}|$ and c_s is the speed of sound which we will assume to be constant in this problem.

1. Prove that the packet's angular momentum $M = k_\theta r$ stays constant as the acoustic packet moves. Here k_θ is the azimuthal component of the wave number $\mathbf{k}$.

2. Prove that the frequency stays constant as the acoustic packet moves.

3. Find the radial component of the wave number as a function of the radius r.

4. Find the "reflection" condition, i.e. a condition for the wave packets approaching the vortex from infinity to turn and to start moving away from the vortex.

5. What happens when the reflection condition is violated? Sketch examples of acoustic rays satisfying and violating the reflection condition.

4.3.10　Kelvin-Helmholtz instability

Consider a plane-parallel shear flow with profile

$$\mathbf{u} = U\hat{\mathbf{x}} \text{ for } y \geq 0 \quad \text{and} \quad \mathbf{u} = -U\hat{\mathbf{x}} \text{ for } y < 0,$$

where $U = \text{const} > 0$; see figure 4.6.

This flow is a stationary solution of the Euler equation. We will study evolution of irrotational perturbations of the specified shear flow.

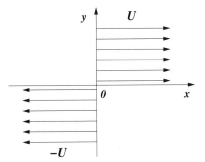

FIGURE 4.6: Kelvin-Helmholtz instability setup: piecewise constant shear flow with a tangential discontinuity.

1. Use Bernoulli's theorem for time-dependent irrotational flow and formulate the pressure boundary conditions at infinity and on on the velocity discontinuity surface. Linearise the resulting equation.

2. Formulate the kinematic boundary conditions on the discontinuity interface corresponding to the top and the bottom side of this surface. Linearise this condition.

3. Write the incompressibility conditions above and below the interface.

4. Consider a plane wave solution

$$\phi_\pm(x, y, t) = A_\pm(y)\, e^{i(kx - \omega t)} + c.c., \quad h = H\, e^{i(kx - \omega t)} + c.c., \quad (4.24)$$

where $A_\pm(y)$ are real functions, H is complex number, k is the perturbation wavenumber and ω is frequency.

Substitute this plane wave solution into the incompressibility conditions and find the shape of the function $A_\pm(y)$.

5. Substitute this plane wave solution into the linearised pressure and the kinematic boundary conditions at the interface which you found in the previous parts. Find ω in terms of k from the resolvability condition of the resulting equations. Show that the resulting values of ω are purely imaginary and one of them has a positive imaginary part, i.e. that the wave perturbations experience an exponential growth (Kelvin-Helmholtz instability).

4.3.11 Rayleigh-Taylor instability

Consider an ideal incompressible fluid at rest subject to gravity in the negative y-direction. The fluid has density ρ_+ in the upper half space $y \geq 0$ and density ρ_- in the lower half space $y < 0$.

This setup is at a steady-state equilibrium which is unstable if $\rho_+ > \rho_-$, i.e. when the upper fluid is heavier than the lower one (e.g. water above oil). We will study such an instability by considering evolution of small irrotational perturbations of the specified configuration.

1. Use Bernoulli's theorem for time-dependent irrotational flow and formulate the pressure boundary condition on the density discontinuity surface. Linearise the resulting equation.

2. Formulate the kinematic boundary conditions on the discontinuity interface corresponding to the top and the bottom side of this surface. Linearise this condition.

3. Write the incompressibility conditions above and below the interface.

4. Consider a plane wave solution

$$\phi_\pm(x, y, t) = A_\pm(y)\, e^{i(kx-\omega t)} + c.c., \quad h = H\, e^{i(kx-\omega t)} + c.c., \quad (4.25)$$

where $A_\pm(y)$ are real functions, H is a complex number, k is the perturbation wavenumber and ω is the frequency.

Substitute this plane wave solution into the incompressibility conditions and find the shape of the function $A_\pm(y)$.

5. Substitute the plane wave solution into the linearised pressure and the kinematic boundary conditions at the interface which you found in the previous parts. Find ω in terms of k from the resolvability condition of the resulting equations. Find the condition under which the values of ω are purely imaginary and one of them has a positive imaginary part, i.e. that the wave perturbations experience an exponential growth (Rayleigh-Taylor instability).

4.3.12 Rapid distortion theory

Consider a plane-parallel shear flow with profile

$$\mathbf{U} = s\, y\, \hat{\mathbf{x}}, \qquad (4.26)$$

where $\hat{\mathbf{x}}$ is the unit vector along the x-axis, $s = \text{const} > 0$ is the shear rate; see figure 4.7.

This flow is a stationary solution of the Euler equation (1.5) (without external forcing, $\mathbf{f} = 0$). We will study the evolution of three-dimensional wave-like perturbations of the specified shear flow. The crucial difference with the above examples of the Kelvin-Helmholtz and the Rayleigh-Taylor instabilities is that in the present case the evolution is "non-modal", i.e. the wave number of the perturbations will no longer be constant, and their growth will not be exponential but algebraic.

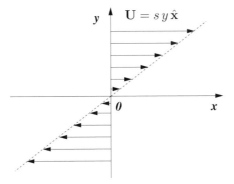

FIGURE 4.7: Constant-shear flow.

1. Consider three-dimensional velocity and pressure perturbations, $\tilde{\mathbf{u}}$ and $\tilde{p}$, of the shear flow (4.26) and linearise the Euler equation with respect to these perturbations.

2. Assume that the perturbations have wave-like shape,

$$\tilde{\mathbf{u}} = \hat{\mathbf{u}}(t)\, e^{i\mathbf{k}(t)\cdot\mathbf{x}} + c.c., \quad \text{and} \quad \tilde{p} = \hat{p}(t)\, e^{i\mathbf{k}(t)\cdot\mathbf{x}} + c.c., \tag{4.27}$$

where $\mathbf{k}(t) = (k_x(t), k_y(t), k_z(t))$ is a 3D wavevector. Substitute (4.27) into the linearised Euler equation and exclude $\hat{p}$ from these equations using the incompressibility condition.

3. Find the time dependence $\mathbf{k}(t)$ for which the coefficients in the linearised Euler equation become independent of $\mathbf{x}$.

4. Assuming that $\mathbf{k}(t)$ has the form found in the previous part, solve the linearised Euler equation and thereby find $\hat{\mathbf{u}}$.

5. Find the asymptotic behaviour of $\hat{\mathbf{u}}$ as a function of t at $t \to +\infty$ assuming that the initial perturbation has a single initial wavenumber $\mathbf{k}(0) = (q_x, q_y, q_z)$.

6. Now assume that the initial perturbation contains a continuous range of wavenumbers with distribution $\hat{\mathbf{u}}_0(\mathbf{q})$ which remains finite when $q_x \to 0$. Find the large-time behaviour of the part of the perturbation with $q_x \sim 1/(st)$.

4.4   Solutions

4.4.1   Model solution to question 4.3.1

1. Substituting $\hat{b}(\mathbf{k}) = B\,\delta(\mathbf{k} - \mathbf{k}_c)$ into (4.11) we have

$$a(\mathbf{x}, t) = B\,e^{i(\mathbf{k}_c \cdot \mathbf{x} - \omega(\mathbf{k}_c)t)} + c.c.,$$

   i.e. we recover the monochromatic wave expression (4.3).

2. If the peak is narrow enough, one can Taylor expand function $\omega(\mathbf{k})$ around $\mathbf{k}_c$, namely

$$\omega(\mathbf{k}) = \omega(\mathbf{k}_c) + (\mathbf{k} - \mathbf{k}_c) \cdot \mathbf{c}_g(\mathbf{k}_c) + o(|\mathbf{k} - \mathbf{k}_c|), \qquad (4.28)$$

   where

$$\mathbf{c}_g = \frac{\partial \omega(\mathbf{k})}{\partial \mathbf{k}}.$$

   The expansion (4.28) is valid if $|\mathbf{k} - \mathbf{k}_c| \ll |\omega(\mathbf{k}_c)|/|\nabla_{\mathbf{k}_c}\omega(\mathbf{k}_c)|$ i.e. the characteristic width σ of function $\hat{b}(\mathbf{k})$ must be $\ll |\omega(\mathbf{k}_c)|/|\mathbf{c}_g(\mathbf{k}_c)|$. Substituting (4.28) into (4.11) we have:

$$a(\mathbf{x}, t) \approx \int_{\mathbf{k} \in \mathbb{R}^d} \hat{b}(\mathbf{k})\, e^{i(\mathbf{k} \cdot \mathbf{x} - \omega(\mathbf{k}_c)t - (\mathbf{k} - \mathbf{k}_c) \cdot \mathbf{c}_g\, t)}\, d\mathbf{k} + c.c. = a_c a_e + c.c., \quad (4.29)$$

   with the monochromatic carrier wave part

$$a_c = e^{i(\mathbf{k}_c \cdot \mathbf{x} - \omega(\mathbf{k}_c)\, t)}$$

   and the envelope part

$$a_e = \int_{\mathbf{k} \in \mathbb{R}^d} \hat{b}(\mathbf{k})\, e^{i(\mathbf{k} - \mathbf{k}_c) \cdot (\mathbf{x} - \mathbf{c}_g\, t)}\, d\mathbf{k}.$$

   Because of the multiplicative factor $\hat{b}(\mathbf{k})$, the contribution to this integral comes only from the wave vectors such that $|\mathbf{k} - \mathbf{k}_c| \sim \sigma \ll |\omega(\mathbf{k}_c)|/|\mathbf{c}_g(\mathbf{k}_c)|$. Thus, the main contribution to a_e comes from $x' = (\mathbf{x} - \mathbf{c}_g\, t) \sim 1/\sigma$: for x' significantly greater than $1/\sigma$ the contributions of different ranges cancel each other because of the oscillations in $e^{i(\mathbf{k} - \mathbf{k}_c) \cdot (\mathbf{x} - \mathbf{c}_g\, t)}$. Therefore, a_e decays at a typical distance $\sim 1/\sigma$ i.e. much slower than the wavelength of the carrier wave part a_c: $\lambda_c = 2\pi/k_c$.

3. Because the $\mathbf{x}$ and t dependence of a_e appears only via the combination $x' = (\mathbf{x} - \mathbf{c}_g\, t)$, the envelope (and therefore the wave packet as a whole) moves with the group velocity $\mathbf{c}_g$ defined above.

4. The $\mathbf{x}$ and t dependence of a_c appears in the combination $(\mathbf{k}_c \cdot \mathbf{x} - \omega(\mathbf{k}_c)\, t) = \mathbf{k}_c \cdot \left(\mathbf{x} - (\omega(\mathbf{k}_c)\mathbf{k}_c/k_c^2)\, t\right)$. Therefore, the wave crests move with the phase velocity $\mathbf{c}_{ph}(\mathbf{k}_c)$ where $\mathbf{c}_{ph}(\mathbf{k}) = \frac{\omega(\mathbf{k})}{|\mathbf{k}|}\frac{\mathbf{k}}{|\mathbf{k}|}$.

5. In the 1D case, the Taylor expansion of $\omega(k)$ around k_c is

$$\omega(k) = \omega(k_c) + (k - k_c)c_g(k_c) + \frac{1}{2}(k - k_c)^2\omega''(k_c) + o(|k - k_c|^2), \quad (4.30)$$

where double prime stands for the second derivative. Then up to the second order in δ for the envelope part we have:

$$a_e = \int_{k \in \mathbb{R}} \hat{b}(k)\, e^{i(k-k_c)(x-c_g\, t) - \frac{i}{2}(k-k_c)^2\omega''(k_c)t}\, dk.$$

As we see, the x and t dependence no longer appears only via the combination $x' = (x - c_g\, t)$, so in addition to the motion of the wave packet as a whole with the group velocity there is an extra t dependence in the envelope a_e if $\omega'' \neq 0$. Thus the condition for the waves to be non-dispersive, i.e. such that their envelope never changes its shape, at this order is $\omega'' = 0$.

4.4.2 Model solution to question 4.3.2

1. For irrotational flow, $\mathbf{u} = \nabla\phi$ (ϕ is the velocity potential), Bernoulli's theorem for time-dependent irrotational flow is

$$\partial_t\phi + \frac{p}{\rho} + \frac{(\nabla\phi)^2}{2} + gz = C, \quad (4.31)$$

where C is a constant (see equation (2.5)).

The pressure boundary condition states that $p = p_0$ on the water surface, which means that at $z = h(x,t)$ equation (4.31) becomes:

$$\left[\partial_t\phi + \frac{(\nabla\phi)^2}{2}\right]_{z=h} + gh = 0, \quad (4.32)$$

where we have chosen the constant $C = \frac{p_0}{\rho}$, which corresponds to a calibration condition that $h = \phi = 0$ in the motionless state.

2. The incompressibility condition is $\nabla \cdot \mathbf{u} = \nabla^2\phi = 0$. Substituting in here $\phi(x, z, t) = B(z)\, e^{i(kx-\omega t)} + c.c.$ we get

$$\frac{d^2 B}{dz^2} - k^2 B = 0.$$

The solution of this equation decaying at $z \to -\infty$ is

$$B(z) = B_0 e^{kz},$$

where B_0 is an arbitrary complex constant (we have chosen $k > 0$).

3. Linearising the pressure boundary condition (4.32) we have

$$[\partial_t \phi]_{z=0} + gh = 0. \qquad (4.33)$$

Linearising the kinematic boundary condition (3.4) we have

$$\partial_t h = u_z = [\partial_z \phi]_{z=0}. \qquad (4.34)$$

Substituting (4.12) into the linearised pressure and the kinematic boundary conditions, (4.33) and (4.34) respectively, we have:

$$-i\omega \frac{H}{2} = kB_0, \qquad (4.35)$$

$$-i\omega B_0 + g\frac{H}{2} = 0. \qquad (4.36)$$

The resolvability condition of this system gives us the expression of the frequency ω:

$$\omega = \pm\sqrt{gk}.$$

Here $k > 0$, and the plus/minus sign corresponds to the wave propagating in the positive/negative x-direction. Alternatively we could have allowed k to be of any sign and chosen $\omega > 0$: $\omega = \sqrt{g|k|}$. In this case the wave would propagate in the direction of k.

Substituting the found relation for ω into (4.36), we have

$$B_0 = -i\sqrt{\frac{g}{k}}\frac{H}{2}.$$

Thus the solution for the velocity potential is:

$$\phi = H\sqrt{\frac{g}{k}}\, e^{kz}\, \sin(kx - \omega t). \qquad (4.37)$$

4. When linearising the kinematic boundary condition, we neglected the nonlinear term $u_x \partial_x h$ assuming it to be small in comparison with the linear term u_z. In our wave solution we have $|u_x| \sim |u_z| \sim kB$. Thus, the linearisation is justified if

$$\partial_x h \ll 1,$$

i.e. if the angle of inclination of the water surface remains small. In terms of the wave amplitude H this condition is

$$kH \ll 1.$$

It is easy to check that neglecting of the other nonlinear terms is valid under the same condition.

5. A particle which in the motionless fluid is located at the Eulerian position (x, z) will move along a Lagrangian path around (x, z):

$$(x(t), z(t)) = (x, z) + (\tilde{x}(t), \tilde{z}(t)).$$

For the particle trajectories $\dot{\tilde{x}}(t) = u_x$, $\dot{\tilde{z}}(t) = u_z$ we have

$$\dot{\tilde{x}}(t) = \partial_x \phi = H\omega\, e^{kz} \cos(kx - \omega t), \tag{4.38}$$
$$\dot{\tilde{z}}(t) = \partial_z \phi = H\omega\, e^{kz} \sin(kx - \omega t), \tag{4.39}$$

where we have neglected the deviation of the particle from it equilibrium position (x, z) in the right-hand sides of the above equations assuming them to be small (which is valid under the same condition as the linearisation procedure).

Equations (4.38) and (4.39) describe motion along the circles

$$\tilde{x}^2 + \tilde{z}^2 = r^2, \quad \text{where} \quad r = He^{kz}.$$

4.4.3 Model solution to question 4.3.3

1. Only pressure gradients and not the pressure itself contribute to the force on a fluid particle. If we added to an existing pressure field a constant at each point of the physical space, then fluid motion would not change. Thus the pressure of air above the free surface cannot influence the dynamical characteristics of the water waves such as, e.g. the wave frequency.

2. The physical dimension of g is L/T^2—as for any acceleration. The physical dimension of k is $1/L$ and the physical dimension of the wave frequency ω is $1/T$. The only possible combination of g, ρ and k which would have the dimension of ω is

$$\omega = \omega(k) = C\sqrt{gk},$$

where C to an order one dimensionless constant (which in this case appears to be 1).

This expression is independent of the density because the latter contains the mass dimension (kg) whereas the rest of the quantities, ω, g and k do not contain the dimension of the mass.

3. The physical dimension of the surface tension coefficient σ, based on its definition as a force per unit length per unit density, is:

$$[\sigma] = [mass][acceleration]/L = [kg]/T^2.$$

4. The unique combination of σ, k and density ρ which would have the dimension of ω is

$$\omega = \omega(k) = C\left(\frac{\sigma}{\rho}\right)^{1/2} k^{3/2},$$

where C to an order one dimensionless constant (which, again, appears to be 1).

5. The crossover wave number k_c at which the capillary effect become more important than the gravity effects may be estimated from balancing the respective expressions for the gravity and the capillary wave frequencies:

$$\sqrt{gk} \sim \left(\frac{\sigma}{\rho}\right)^{1/2} k^{3/2} \quad \rightarrow \quad k_c = \sqrt{\frac{g\rho}{\sigma}}.$$

Using the numbers $\sigma = 0.072\ N/m$, $\rho = 1000\ kg/m^3$ and $9.8\ m/s^2$, for the crossover wave length λ_c we have:

$$\lambda_c = \frac{2\pi}{k_c} = 2\pi\sqrt{\frac{\sigma}{g\rho}} = 0.017\ m = 1.7\ cm.$$

4.4.4   Model solution to question 4.3.4

1. The rigid body rotation velocity is $\mathbf{U} = (\mathbf{\Omega} \times \mathbf{r})$, so the vorticity is $2\mathbf{\Omega}$.

2. Comparing $(\mathbf{u}\cdot\nabla)\mathbf{u}$ and $2(\mathbf{\Omega}\cdot\mathbf{u})$: nonlinear term $(\mathbf{u}\cdot\nabla)\mathbf{u}$ can be neglected if $u \ll \Omega/k$.

3. Taking curl of the Euler equation (1.5) (without external forcing, $\mathbf{f} = 0$), we have

$$\partial_t \nabla \times \mathbf{u} = 2\Omega\partial_z\mathbf{u}.$$

Substituting the plane wave expression, we have

$$\omega(\mathbf{k} \times \mathbf{A}) = 2i\Omega k_z \mathbf{A}. \tag{4.40}$$

After cross multiplication of both sides by $\mathbf{k}$:

$$-\omega k^2 \mathbf{A} = 2i\Omega k_z (\mathbf{k} \times \mathbf{A}). \tag{4.41}$$

Comparing the equations (4.40) and (4.41) we have

$$\omega = \pm 2\Omega\frac{k_z}{k}.$$

4. Substituting the above expression for the frequency into the equation (4.40), we have

$$A_{\parallel} = 0, \quad \frac{1}{k}(\mathbf{k} \times \mathbf{A}_{\perp}) = \pm i\mathbf{A}_{\perp}, \tag{4.42}$$

where $A_{\parallel}$ and $\mathbf{A}_{\perp}$ are the parallel and the perpendicular to $\mathbf{k}$ the projections of $\mathbf{A}$ respectively. Thus, the perpendicular to $\mathbf{k}$ components of $\mathbf{A}$ oscillate with the same amplitude and with phase difference $\pi/2$ with respect to each other (since $i = e^{\pi/2}$), which means that the wave polarisation is circular.

5. For the group velocity we have

$$\mathbf{c}_g = \partial_{\mathbf{k}}\omega = \pm\frac{2\Omega}{k^3}(-k_zk_x, -k_zk_y, k_\perp^2). \tag{4.43}$$

The waves are dispersive: the group velocity depends on $k = |\mathbf{k}|$ for any fixed direction of $\mathbf{k}$. The waves are anisotropic: ω depends on the direction of $\mathbf{k}$ for any fixed $k = |\mathbf{k}|$.

From the equation (4.43) we have $\mathbf{c}_g \cdot \mathbf{k} = 0$. Thus the group velocity is perpendicular to the phase velocity, i.e. the inertial wave packets propagate along their wave crests (rather than perpendicular to them an in the isotropic wave cases); c.f. similar result for the internal waves.

4.4.5 Model solution to question 4.3.5

1. Integrating the equation (4.13) at rest we have

$$p_0(z) = gh\rho_0(z).$$

2. Linearising the equations (4.13) and (4.15) we have

$$\partial_t\tilde{\mathbf{u}} = -\frac{1}{\rho_0}\nabla\tilde{p} + \frac{\tilde{\rho}}{\rho_0^2}p_0'\hat{\mathbf{z}}, \tag{4.44}$$

and

$$\partial_t\tilde{\rho} + \tilde{u}_z\rho_0' = 0. \tag{4.45}$$

For the incompressibility condition we have

$$\nabla \cdot \tilde{\mathbf{u}} = 0. \tag{4.46}$$

Substituting (4.16), (4.17) and (4.18) and the profiles $\rho_0(z)$ and $p_0(z)$ into the above equations, we have

$$-i\omega\mathbf{A}_\perp = -i\mathbf{k}_\perp P, \tag{4.47}$$

$$-i\omega A_z = (-ik_z + \frac{1}{2h})P - gR, \tag{4.48}$$

$$\tag{4.49}$$

$$-i\omega R = \frac{1}{h}A_z \tag{4.50}$$

and

$$(ik_z + \frac{1}{2h})A_z + i\mathbf{k}_\perp \cdot \mathbf{A}_\perp = 0. \tag{4.51}$$

Here $\perp$ denotes the vector projection on the horizontal plane (x, y).

Then we obtain the dispersion relation by reducing this system in the following sequence:

- Scalar-multiply equation (4.47) by $\mathbf{k}_\perp$ and solve it for P.

- Use the resulting expression for P and (4.51) to obtain an expression for P in terms of A_z and substitute it into equation (4.48).

- Solve equation (4.50) for R and substitute the result into equation (4.47).

- The only remaining variable A_z will cancel out from the resulting equation, which yields the systems solvability condition, i.e. the dispersion relation:

$$\omega = \omega(\mathbf{k}) = \frac{Nk_\perp}{\sqrt{k^2 + \frac{1}{4h}}}, \qquad (4.52)$$

  where $N = \sqrt{g/h}$ is the buoyancy frequency.

3. In the limit $kh \gg 1$ we have $\omega(\mathbf{k}) = Nk_\perp/k$. For the group velocity we find

$$\mathbf{c}_g = \partial_\mathbf{k}\omega = \frac{N}{k^3 k_\perp}(k_x k_z^2, k_y k_z^2, -k_z k_\perp^2). \qquad (4.53)$$

The waves are dispersive as the group velocity depends on k. The waves are anisotropic as ω depends on the direction of $\mathbf{k}$ (i.e. not only on $k = |\mathbf{k}|$).

From equation (4.53) we have $\mathbf{c}_g \cdot \mathbf{k} = 0$. Thus the group velocity is perpendicular to the phase velocity, i.e. the inertial wave packets propagate along their wave crests (rather than perpendicular to them as in the isotropic wave cases); c.f. similar result for the inertial waves.

4.4.6   Model solution to question 4.3.6

1. Substituting $\rho = \rho_0 + \tilde{\rho}$, $p = p_0 + \tilde{p}$ and $\mathbf{u} = \tilde{\mathbf{u}}$ into equations (4.19), (4.20) and (4.21) and linearising, we have

$$\partial_t \tilde{\mathbf{u}} = -\frac{1}{\rho_0}\nabla\tilde{p} = -\frac{c_s^2}{\rho_0}\nabla\tilde{\rho}, \qquad (4.54)$$

where $c_s = \sqrt{\partial_{\rho_0} p_0} = \sqrt{\gamma p_0/\rho_0}$ is the speed of sound, and

$$\partial_t \tilde{\rho} + \rho_0 \nabla \cdot \tilde{\mathbf{u}} = 0. \qquad (4.55)$$

2. Substituting the harmonic wave perturbations (4.22) and (4.23) into the linearised equations, we have:

$$-\omega\mathbf{A} = -\frac{c_s^2}{\rho_0}\mathbf{k}B, \qquad (4.56)$$

and

$$-\omega B + \rho_0 \, \mathbf{k} \cdot \mathbf{A} = 0. \tag{4.57}$$

From the solvability condition of these equations we find the sound-wave dispersion relation,

$$\omega = \omega(\mathbf{k}) = \pm c_s k,$$

where $k = |\mathbf{k}|$. Using this expression, we get the relation between $\mathbf{A}$ and B:

$$\mathbf{A} = \frac{c_s \mathbf{k}}{\rho_0 k} B.$$

3. In the linear approximation we have assumed smallness of the nonlinear term $(\mathbf{u} \cdot \nabla)\mathbf{u}$ with respect to $\partial_t \mathbf{u}$. In terms of the amplitude $\mathbf{A}$ this means $|\mathbf{A}| \ll c_s$. Easy to check that the other nonlinear terms are small under the same condition.

4.4.7 Model solution to question 4.3.7

1. From the x-component of the ray tracing equation (4.10) we have

$$\dot{k}_x = -\partial_x \omega = 0.$$

Therefore k_x stays constant as the acoustic packet moves.

2. The time derivative of the frequency along the wave packet's trajectory is

$$D_t \omega = \partial_t \omega + \dot{\mathbf{x}}(t) \cdot \nabla \omega + \dot{\mathbf{k}}(t) \cdot \partial_{\mathbf{k}} \omega.$$

Taking into account that $\partial_t \omega = 0$ and substituting $\dot{\mathbf{x}}$ and $\dot{\mathbf{k}}$ from the ray tracing equations (4.9) and (4.10), we have

$$D_t \omega = 0,$$

i.e. the frequency stays constant as the acoustic packet moves.

3. We have

$$\omega = c_s k + \mathbf{k} \cdot \mathbf{u} = c_s \sqrt{k_x^2 + k_y^2} + k_x u(y).$$

Solving for k_y, we have

$$k_y = \pm \sqrt{\frac{(\omega - k_x u(y))^2}{c_s^2} - k_x^2}.$$

4. According to the equation (4.9) the y-component of the packet's velocity is $c_s k_y / k$, so reflections occur when this velocity component is zero, i.e. when $k_y = 0$. This happens at $y = y_r$ where

$$u(y_r) = u_r = \frac{\omega}{k_x} - c_s,$$

which gives

$$y_r = \frac{1}{\delta} \tan \frac{\pi u_r}{2u_0} = \frac{1}{\delta} \tan \frac{\pi(\omega/k_x - c_s)}{2u_0}.$$

Since y_r must be positive, the wave packet's reflection may only occur if $k_x > 0$. The second condition for the wave packet's reflection is that $u_r < \max u(y) = u_0$, which gives

$$\frac{k_0}{k_x} - 1 < \frac{u_0}{c_s},$$

where $k_0 = \sqrt{k_x^2 - k_{0y}^2}$ is the absolute value of the initial wave number.

5. A sketch of the acoustic rays is given in figure 4.8. Rays directed along the wind are bent toward the ground and the most oblique of them get reflected. This explains amplification of sound on the downwind side of the source. Rays directed against the wind are bent away from the ground including the ray with $k_{0y} = 0$; hence appearance of a quiet zone upwind of the source.

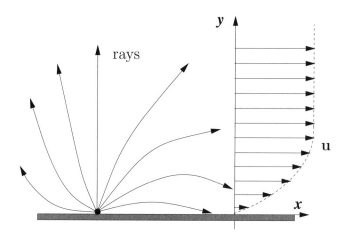

FIGURE 4.8: Sound rays in a shear flow.

4.4.8 Model solution to question 4.3.8

1. From the x-component of the ray tracing equation (4.10) we have

$$\dot{k}_x = -\partial_x \omega = 0.$$

Therefore k_x stays constant as the acoustic packet moves. The same is true for the z-component of the wave vector. (Actually, one can choose the y- and z-axes so that the initial k_z is zero, in which case it will remain zero at all time).

2. The time derivative of the frequency along the wave packet's trajectory is

$$D_t\omega = \partial_t\omega + \dot{\mathbf{x}}(t) \cdot \nabla\omega + \dot{\mathbf{k}}(t) \cdot \partial_{\mathbf{k}}\omega.$$

Taking into account that $\partial_t\omega = 0$ and substituting $\dot{\mathbf{x}}$ and $\dot{\mathbf{k}}$ from the ray tracing equations (4.9) and (4.10), we have

$$D_t\omega = 0,$$

i.e. the frequency stays constant as the acoustic packet moves.

3. We have

$$\omega = c_s(y)k = c_s(y)\sqrt{k_\perp^2 + k_y^2}.$$

Solving for k_y, we have

$$k_y(y) = \pm\sqrt{\frac{\omega^2}{c_s^2(y)} - k_\perp^2}.$$

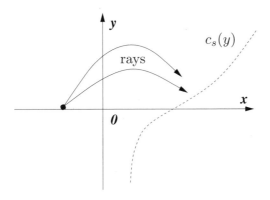

FIGURE 4.9: Sound rays in a stratified flow.

4. According to equation (4.9) the y-component of the packet's velocity is $c_s k_y/k$, so the reflections occur when this velocity component is zero, i.e. when $k_y = 0$. This happens at $y = y_r$ where

$$c_s(y_r) = \frac{c_s(0)k_0}{k_\perp} = \frac{c_s(0)}{\sin\alpha}, \tag{4.58}$$

where α is the angle between the initial wave vector and the y-axis.

5. A sketch of acoustic rays is shown in the figure 4.9. A sketch of rays in a system which has a speed of sound minimum around some layer y_m is shown in the figure 4.10. The rays are trapped in some slab: they move in a zigzag fashion getting repeatedly reflected at the regions with higher values of the speed of sound. Therefore, systems with a minimum of the speed of sound have a wave guiding property.

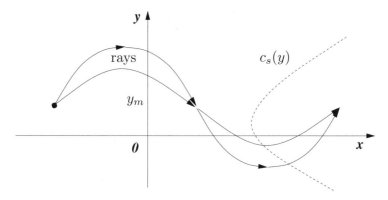

FIGURE 4.10: Sound rays in a stratified flow.

6. From relation (4.58) we know that the reflection will take place if $c_2 > c_s(y_r)$ i.e. if

$$c_1/c_2 < \sin \alpha.$$

If this condition is violated, then the wave packet will not reflect—it will emerge at the other side of the interface (refraction). The relation between the angles of incidence and refraction can be found from the frequency conservation:

$$c_1 k_1 = c_2 k_2,$$

which, together with $k_{1\perp} = k_{2\perp}$, gives the Snell's law for the angles of incidence and refraction:

$$\sin \alpha_2 = \frac{c_2}{c_1} \sin \alpha_1.$$

4.4.9 Model solution to question 4.3.9

1. Note that the wave packet's angular momentum $M = k_\theta r$ is the z-component of $(\mathbf{k} \times \mathbf{r})$. We have:

$$D_t(\mathbf{k} \times \mathbf{r}) = (\mathbf{k} \times (c_s\mathbf{k}/k + \mathbf{u})) + (\mathbf{r} \times \nabla\omega) = (\mathbf{k} \times \mathbf{u}) + (\mathbf{r} \times \nabla(\mathbf{k} \cdot \mathbf{u})).$$

Now $\mathbf{u} = \frac{u_\theta}{r}(\mathbf{z} \times \mathbf{r})$, so

$$(\mathbf{r} \times \nabla(\mathbf{k} \cdot \mathbf{u})) = (\mathbf{r} \times \nabla(\frac{u_\theta}{r}\mathbf{k} \cdot (\mathbf{z} \times \mathbf{r}))) = -\frac{u_\theta}{r}(\mathbf{r} \times \nabla(\mathbf{k} \times \mathbf{r})_z) =$$

$$-\frac{u_\theta}{r}(\mathbf{k} \cdot \mathbf{r}) = -(\mathbf{k} \times \mathbf{u}).$$

Thus $D_t M = 0$: the angular momentum is conserved.

2. The time derivative of the frequency along the wave packet's trajectory is

$$D_t\omega = \partial_t\omega + \dot{\mathbf{x}}(t) \cdot \nabla\omega + \dot{\mathbf{k}}(t) \cdot \partial_\mathbf{k}\omega.$$

Taking into account that $\partial_t \omega = 0$ and substituting $\dot{\mathbf{x}}$ and $\dot{\mathbf{k}}$ from the ray tracing equations (4.9) and (4.10), we have

$$D_t \omega = 0,$$

i.e. the frequency stays constant as the acoustic packet moves.

3. The frequency is

$$\omega = c_s \sqrt{k_r^2 + k_\theta^2} + u_\theta k_\theta.$$

Solving for k_r^2 we have

$$k_r^2 = \frac{1}{c_s^2}(\omega - u_\theta k_\theta)^2 - k_\theta^2 = \frac{1}{c_s^2}\left(\omega - \frac{\Gamma M}{2\pi r^2}\right)^2 - \frac{M^2}{r^2}.$$

4. Reflection occurs when $k_r = 0$ i.e. when

$$\omega = c_s |k_\theta| + u_\theta k_\theta = \frac{c_s |M|}{r} + \frac{\Gamma M}{2\pi r^2}.$$

The reflection point $r = r_*$ is given by a positive solution of this quadratic equation

$$r_* = \frac{c_s |M| + \sqrt{c_s^2 M^2 + 2\Gamma M \omega / \pi}}{2\omega}.$$

The solution for r_* exists if the discriminant is positive, which occurs when $M \geq 0$ or $M \leq -\frac{2\Gamma \omega}{\pi c_s^2}$.

If the wave packet is approaching the vortex from infinity with starting absolute value of the wave number k_∞ and impact parameter a (see the figure 4.11), then $\omega = c_s k_\infty$ and $M = -k_\infty a$. In this case the reflection condition is

$$a \leq 0 \quad \text{or} \quad a \geq \frac{2\Gamma}{\pi c_s}.$$

5. When the reflection condition is violated, i.e. if $0 < a < \frac{2\Gamma}{\pi c_s}$, the wave packet does not turn back: it falls onto the vortex. A sketch of examples of acoustic rays satisfying and violating the reflection condition is given in figure 4.11.

4.4.10 Model solution to question 4.3.10

1. For irrotational flow, $\mathbf{u}_\pm = \pm U\hat{\mathbf{x}} + \nabla \phi_\pm$ ($\phi_\pm$ is the velocity potential of the perturbations), Bernoulli's theorem for time-dependent irrotational flow is

$$\partial_t \phi_\pm + \frac{p_\pm}{\rho} + \frac{(\pm U\hat{\mathbf{x}} + \nabla \phi_\pm)^2}{2} = C_\pm, \tag{4.59}$$

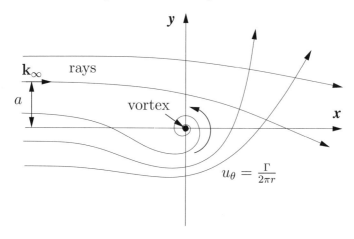

FIGURE 4.11: Sound rays in a point vortex flow.

where $C_\pm$ is a constant (see equation (2.5)).

The pressure boundary condition states that $p_\pm = p_0$ at $y \to \pm\infty$. Therefore, we choose the constant in Bernoulli's equation $C_\pm = \frac{p_0}{\rho} + \frac{U^2}{2}$, which corresponds to a calibration condition that $\phi_\pm \to 0$ for $y \to \pm\infty$.

Linearising the equation (4.59) we have

$$\partial_t \phi_\pm + \frac{p_\pm}{\rho} \pm U \partial_x \phi_\pm = \frac{p_0}{\rho}. \tag{4.60}$$

At the interface the pressures p_+ and p_- must match. In the linear approximation we can ignore deviation of the interface $y = h(x,t)$ from its unperturbed location, $y = 0$. Thus, using the equation (4.60) and matching $p_+ = p_-$ at $y = 0$, we have

$$\partial_t \phi_+ + U \partial_x \phi_+ = \partial_t \phi_- - U \partial_x \phi_-. \tag{4.61}$$

2. The kinematic boundary conditions on the interface corresponding to the top and the bottom side of the interface are:

$$\partial_t h + (u_x)_\pm \partial_x h = (u_y)_\pm \quad \text{at} \quad y = h(x,t). \tag{4.62}$$

Linearising this equation we have

$$\partial_t h \pm U \partial_x h = \partial_y \phi_\pm \quad \text{at} \quad y = 0. \tag{4.63}$$

3. For the incompressibility conditions we have:

$$\nabla^2 \phi_\pm = 0. \tag{4.64}$$

4. Substituting the plane wave solution (4.24) into the equation (4.75) we have

$$A_\pm(y) = B_\pm e^{\mp ky},$$

where $B_\pm$ are real constants. Here we assume that $k > 0$ and choose the solution that decays at $y \to \pm\infty$.

5. Substituting the plane wave solution (4.24) into the equation (4.61) we have

$$-\omega B_+ + Uk B_+ = -\omega B_- - Uk B_-, \qquad (4.65)$$

and from the linearised kinematic boundary conditions (4.63) we have:

$$-i\omega H = -UikH - kB_+ = UikH + kB_-. \qquad (4.66)$$

$$H = \frac{kB_-}{i(-\omega - Uk)} = \frac{kB_+}{i(\omega - Uk)}. \qquad (4.67)$$

From the equations (4.65) and (4.67) we have the following resolvability condition:

$$(\omega + Uk)^2 = -(\omega - Uk)^2. \qquad (4.68)$$

or

$$\omega = i\gamma, \quad \gamma = \pm Uk. \qquad (4.69)$$

Thus we see that there are two modes with purely imaginary frequencies, one with a positive and another with a negative imaginary parts. We have

$$e^{-i\omega t} = e^{\gamma t},$$

i.e. the mode with positive γ is unstable.

4.4.11 Model solution to question 4.3.11

1. For irrotational flow, $\mathbf{u}_\pm = \nabla\phi_\pm$ ($\phi_\pm$ is the velocity potential of the perturbations), Bernoulli's theorem for time-dependent irrotational flow is

$$\partial_t \phi_\pm + \frac{p_\pm}{\rho_\pm} + \frac{(\nabla\phi_\pm)^2}{2} - gy = C_\pm, \qquad (4.70)$$

where $C_\pm$ is a constant (see equation (2.5)).

We choose the constant in Bernoulli's equation to be $C_\pm = \frac{p_0}{\rho_\pm}$, where p_0 is a reference pressure at the density discontinuity surface when the latter is at rest.

The pressure boundary condition states that $p_+ = p_-$ at the moving density discontinuity interface $h = y(x,t)$.

Linearising the equation (4.70) we have

$$\partial_t \phi_\pm + \frac{p_\pm - p_0}{\rho_\pm} - gy = 0. \tag{4.71}$$

At the interface the pressures p_+ and p_- must match. In the linear approximation we can ignore deviation of the interface $y = h(x,t)$ from its unperturbed location, $y = 0$, in all terms except for $gy = gh(x,t)$. Thus, using the equation (4.71) and matching $p_+ = p_-$ at $y = 0$, we have

$$\rho_+(\partial_t \phi_+ - gh) = \rho_-(\partial_t \phi_- - gh). \tag{4.72}$$

2. The kinematic boundary conditions on the interface corresponding to the top and the bottom side of the interface are:

$$\partial_t h + (u_x)_\pm \partial_x h = (u_y)_\pm \ \text{ at } \ y = h(x,t). \tag{4.73}$$

Linearising this equation we have

$$\partial_t h = \partial_y \phi_\pm \ \text{ at } \ y = 0. \tag{4.74}$$

3. For the incompressibility conditions we have:

$$\nabla^2 \phi_\pm = 0. \tag{4.75}$$

4. Substituting the plane wave solution (4.25) into the equation (4.75) we have

$$A_\pm(y) = B_\pm e^{\mp ky},$$

where $B_\pm$ are real constants. He we assume that $k > 0$ and choose the solution that decays at $y \to \pm\infty$.

5. Substituting the plane wave solution (4.25) into the equation (4.72) we have

$$\rho_+(-i\omega B_+ - gH) = \rho_-(-i\omega B_- - gH), \tag{4.76}$$

and from the linearised kinematic boundary conditions (4.74) we have:

$$-i\omega H = -kB_+ = kB_-. \tag{4.77}$$

From the equations (4.76) and (4.77) we have the following resolvability condition:

$$\rho_+(\omega^2/k - g) = \rho_-(-\omega^2/k - g), \tag{4.78}$$

or

$$\omega = \sqrt{\frac{(\rho_- - \rho_+)gk}{\rho_- + \rho_+}}. \tag{4.79}$$

Thus we see that in the case $\rho_- > \rho_+$ we have a real frequency which corresponds to a propagating wave. This is a generalisation of the surface gravity wave considered in question 4.3.2.

In the case $\rho_- < \rho_+$ there are two modes with purely imaginary frequencies, $\omega = i\gamma$, one with a positive and another with a negative imaginary parts:

$$\gamma = \pm\sqrt{\frac{(\rho_+ - \rho_-)gk}{\rho_- + \rho_+}}. \qquad (4.80)$$

We have

$$e^{i\omega t} = e^{\gamma t},$$

i.e. the mode with positive γ is unstable.

4.4.12 Model solution to question 4.3.12

(a) Consider the three-dimensional velocity and pressure perturbations, $\tilde{\mathbf{u}}$ and $\tilde{p}$, of the shear flow (4.26). The linearised Euler equation is

$$\partial_t \tilde{\mathbf{u}} + (\mathbf{U} \cdot \nabla)\tilde{\mathbf{u}} + (\tilde{\mathbf{u}} \cdot \nabla)\mathbf{U} = -\frac{1}{\rho}\nabla\tilde{p}, \qquad (4.81)$$

or, substituting for $\mathbf{U}$,

$$\partial_t \tilde{\mathbf{u}} + sy\,\partial_x \tilde{\mathbf{u}} + s\tilde{u}_y\,\hat{\mathbf{x}} = -\frac{1}{\rho}\nabla\tilde{p}. \qquad (4.82)$$

Taking divergence of this equation and taking into account the incompressibility condition $\nabla \cdot \tilde{\mathbf{u}} = 0$, we have

$$-\frac{1}{\rho}\nabla^2 \tilde{p} = 2s\,\partial_x \tilde{u}_y. \qquad (4.83)$$

(b) Assuming that the perturbations have wave-like shape (4.27), equations (4.82) and (4.83) become:

$$\partial_t \hat{\mathbf{u}} + i(\dot{\mathbf{k}} \cdot \mathbf{x})\hat{\mathbf{u}} + sy\,ik_x\hat{\mathbf{u}} + s\hat{u}_y\,\hat{\mathbf{x}} = -\frac{1}{\rho}ik\hat{p}. \qquad (4.84)$$

and

$$k^2 \frac{\hat{p}}{\rho} = 2si\,k_x\,\hat{u}_y.$$

Solving the latter expression for $\hat{p}$ and substituting the result in (4.84) we have

$$\partial_t \hat{\mathbf{u}} + i(\dot{\mathbf{k}} \cdot \mathbf{x})\hat{\mathbf{u}} + sy\,ik_x\hat{\mathbf{u}} + s\hat{u}_y\left(\hat{\mathbf{x}} - \mathbf{k}\frac{2k_x}{k^2}\right) = 0. \qquad (4.85)$$

(c) For the coefficients in the linearised Euler equation (4.85) to become independent of $\mathbf{x}$ we must have

$$(\dot{\mathbf{k}} \cdot \mathbf{x}) = -sy\,k_x,$$

or

$$\dot{k}_x = \dot{k}_z = 0 \quad \text{and} \quad \dot{k}_y = -s\,k_x.$$

Solving these equations, we have

$$k_x = q_x, \quad k_z = q_z \quad \text{and} \quad k_y = q_y - sq_x t,$$

where $\mathbf{q} = \mathbf{k}(0)$ is the initial wave number.

(d) Substituting these expressions into (4.85) we have

$$\partial_t \hat{u}_x + s\hat{u}_y \left(1 - \frac{2q_x^2}{q_x^2 + (q_y - sq_x t)^2 + q_z^2}\right) = 0, \quad (4.86)$$

$$\partial_t \hat{u}_y - 2s\hat{u}_y \frac{q_x(q_y - sq_x t)}{q_x^2 + (q_y - sq_x t)^2 + q_z^2} = 0, \quad (4.87)$$

$$\partial_t \hat{u}_z - 2s\hat{u}_y \frac{q_x q_z}{q_x^2 + (q_y - sq_x t)^2 + q_z^2} = 0. \quad (4.88)$$

Integrating the equation (4.87) we have

$$\hat{u}_y(t) = \hat{u}_y(0) \frac{q^2}{q_x^2 + (q_y - sq_x t)^2 + q_z^2}, \quad (4.89)$$

where $q^2 = q_x^2 + q_y^2 + q_z^2$. Substituting (4.89) into (4.86) and (4.88) and integrating, we have

$$\hat{u}_x(t) = \hat{u}_y(0) \frac{q^2}{q_x^2 + q_z^2} \left[-q_z^2 \lambda + \frac{q_x^2}{q^2} \mu\right] + \hat{u}_x(0), \quad (4.90)$$

$$\hat{u}_z(t) = \hat{u}_y(0) \frac{q_x q_z}{q_x^2 + q_z^2} \left[q^2 \lambda + \mu\right] + \hat{u}_z(0), \quad (4.91)$$

where

$$\lambda = \frac{1}{q_x(q_x^2 + q_z^2)^{1/2}} \arctan \frac{stq_x(q_x^2 + q_z^2)^{1/2}}{q^2 - stq_x q_y}, \quad (4.92)$$

$$\mu = \frac{st(q^2 - 2q_y^2 + stq_x q_y)}{q_x^2 + (q_y - sq_x t)^2 + q_z^2}. \quad (4.93)$$

(e) At large time, $t \to +\infty$, we have $\hat{u}_y \to 0$. Also, λ and μ tend to some constant values, and so do $\hat{u}_x$ and $\hat{u}_z$. Thus, there is no instability if only a single wave was chosen as an initial condition.

(f) If the initial perturbation contains a continuous range of wavenumbers with distribution $\hat{\mathbf{u}}_0(\mathbf{q})$ which remains finite when $q_x \to 0$, then at large t the part of the perturbation with $q_x \sim 1/(st)$ will behave as

$$\hat{u}_x \sim st, \quad \hat{u}_y \sim \text{const}, \quad \hat{u}_z \sim \text{const}.$$

Thus we see that such perturbations of the constant shear flow grow algebraically (linearly in time). One can show that the total energy of the perturbations will also grow algebraically,

$$E = \frac{1}{2} \int \hat{\mathbf{u}}^2(\mathbf{k}) \, d\mathbf{k} \approx \frac{1}{2} \int \hat{u}_x^2(\mathbf{k}) \, d\mathbf{k} \sim t.$$

This is because the $\mathbf{k}$-space volume contributing to the integral is proportional to the range of contributing $k_x = q_x$ which shrinks in time as $\sim 1/(st)$.

Chapter 5

Boundary layers

5.1 Background theory

Boundary layers are flows formed in the close vicinity of solid boundaries. Their properties significantly differ from the properties of the flows further away from the boundaries; see figure 5.1. Boundary layers are most clearly defined for laminar flows with high Reynolds numbers, because in this case this will be the only part of the flow where viscosity is important, whereas the flow in the exterior is nearly ideal. One could roughly think that such an ideal flow is slowly varying (often irrotational) and satisfying the free-slip boundary conditions on the solid surface. In the other words, when finding the exterior flow, presence of the boundary layer can be ignored because it is much thinner than the typical length scale L in the exterior flow, $\delta \ll L$; see figure 5.1. On the other hand, the value of the velocity slip serves as a boundary condition at the top side of the thin boundary layer. The role of the viscosity here is to eliminate the free slip by slowing the velocity down to zero rapidly within a thin layer so that the no-slip boundary condition could be satisfied at the bottom side of the boundary layer. Since the viscosity coefficient is small, the velocity gradient in the boundary layer must be large so that the viscosity term would become sufficient to balance the other terms in the Navier-Stokes equation. For this reason the boundary layer is thin— thinner than the curvature radius of the solid surface–and it can be locally thought as a plane-parallel shear flow. These ideas were originally presented in Prandtl's 1905 paper [20] and they remain a conceptual basis for the modern boundary layer theory in laminar flows. Very good detailed presentation of the laminar boundary layer theory can be found in the Acheson's book *Elementary Fluid Dynamics* [1] as well as in *Essentials of Fluid Dynamics* by L. Prandtl [21].

If the flow far from the boundary is turbulent then the structure of the boundary layer is more complicated, and it cannot be defined as a layer beyond which the flow is ideal (and in some cases irrotational). Indeed, turbulent motions span over a wide range of scales including such small scales at which viscosity is important. Also, turbulent flows are never irrotational. In the present chapter we will only consider laminar flows. We will study the

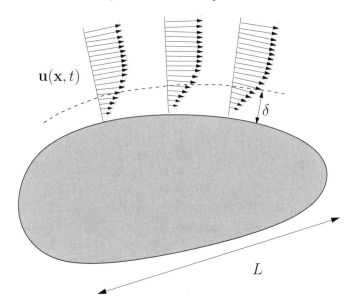

FIGURE 5.1: Boundary layer profile.

structure of a turbulent flow near a solid boundary later when we consider turbulence in chapter 8; see problem 8.3.6.

Below, we will start with a problem which takes us though derivation of a simplified set of equations for the boundary layer flows which takes into account the difference of scales in the boundary layer and in the exterior flow, $\delta \ll L$. These equations will be used further in questions dealing with boundary layers produced by a uniform flow over a flat plate and by a constant strain flow over a flat plate. This will be followed by considering boundary layers produced by an oscillating flow and by a rotating fluid over a rotating boundary.

5.2 Problems

5.2.1 The boundary layer equations

Following the ideas explained in the background theory section above, and in Prandtl's 1905 paper [20], we will consider a laminar high Reynolds number flow over a smooth boundary, and will assume a scale separation. Namely, we will suppose that locally the boundary layer flow is close to being plane parallel, with dominant velocity component along the boundary, and the fastest velocity gradients in the normal to the boundary direction. Such a scale separation will allow us (following Prandtl's 1905 paper) to derive a simplified set of the boundary layer equations.

Consider a flow with a typical velocity U near a bluff body with a characteristic length scale L as in figure 5.1. The flow is laminar and the Reynolds number is high, $Re = UL/\nu \gg 1$, where ν is the viscosity coefficient. The boundary layer is thin with characteristic thickness $\delta \ll L$ and it is attached to the surface, i.e. there are no streamlines originating at the surface and penetrating to distances $\sim L$ away from the surface.

1. Taking into account that the boundary layer flow is locally almost plane-parallel, write a physical estimate for the boundary layer thickness δ from balancing the inertial and the viscous forces within the boundary layer (so that the flow could be slowed down to satisfy the no-slip boundary condition at the solid surface). Express your answer in terms of L and Re.

2. Rewrite the two-dimensional Navier-Stokes equations in terms of the non-dimensional scaled variables

$$x' = \frac{x}{L}, \quad y' = \frac{y}{\delta}, \quad u' = \frac{u}{U}, \quad v' = \frac{v}{(U\delta/L)}, \quad p' = \frac{p}{\rho U^2},$$

where u, v and x, y are the parallel and the perpendicular to the boundary velocity and coordinate components respectively.

3. By taking the limit $Re \to \infty$ with all primed functions and their derivatives with respect to the primed coordinates kept $O(1)$ with respect to $1/Re$, derive the boundary layer equations:

$$u'\frac{\partial u'}{\partial x'} + v'\frac{\partial u'}{\partial y'} = -\frac{\partial p'}{\partial x'} + \frac{\partial^2 u'}{\partial y'^2}, \tag{5.1}$$

$$\frac{\partial p'}{\partial y'} = 0, \tag{5.2}$$

$$\frac{\partial u'}{\partial x'} + \frac{\partial v'}{\partial y'} = 0. \tag{5.3}$$

4. Rewrite the boundary layer equations in terms of the original (unprimed) variables.

5. Formulate the boundary conditions for the obtained equations at $y = 0$ and at $y = \infty$.

5.2.2   A boundary layer over a semi-infinite plate

Consider a boundary layer at a semi-infinite flat plate at $y = 0$, $x \geq 0$ produced by the flow which is in the inviscid exterior (i.e. outside of the boundary layer) has uniform velocity with magnitude U and directed parallel to the plate along the x-axis; see figure 5.2.

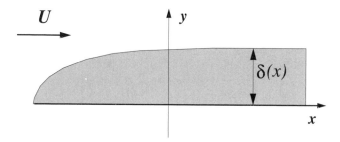

FIGURE 5.2: Boundary layer over a semi-infinite plate.

1. Use a dimensional argument to find δ in terms of ν, U and x.

2. Explain why we should be seeking solution for the velocity components in the boundary layer in the form

$$u = Uf(\eta), \quad \text{and} \quad v = \frac{U\delta}{x}g(\eta).$$

   Find the boundary conditions for $f(\eta)$ and $g(\eta)$.

3. Use the incompressibility condition to find an equation relating $f(\eta)$ and $g(\eta)$.

4. The x-momentum equation in the boundary layer is (c.f. problem 5.2.1):

$$u\partial_x u + v\partial_y u = -\frac{1}{\rho}\partial_x p + \nu\partial_{yy}u, \tag{5.4}$$

   where $p \equiv p(x) = p^{ext}(x,0)$ is the value of the pressure in the external ideal flow at the boundary. Find the first term on the right-hand side of equation (5.4).

   Substitute u and v in terms of f and g into equation (5.4) and find the second equation relating $f(\eta)$ and $g(\eta)$. Using the two equations for $f(\eta)$ and $g(\eta)$, find a single ordinary differential equation for $f(\eta)$. How many boundary conditions does one need to solve this equation? Specify all these boundary conditions.

5.2.3   Boundary layer produced by a pure strain flow

Consider a boundary layer at an infinite flat plate at $y = 0$ produced by a flow which is in the inviscid exterior (i.e. outside of the boundary layer) that has the form of a pure strain flow,

$$\mathbf{u}^{ext} = (u^{ext}, v^{ext}) = (\alpha x, -\alpha y); \tag{5.5}$$

see figure 5.5.

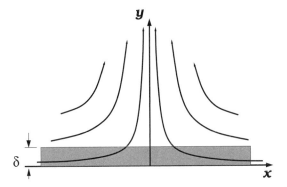

FIGURE 5.3: Boundary layer near a stagnation point.

1. Find the pressure field $p^{ext}(x, y)$ in the inviscid exterior flow using Bernoulli's theorem.

2. The boundary layer equations are (c.f. problem 5.2.1):

$$u\partial_x u + v\partial_y u \;=\; -\frac{1}{\rho}\partial_x p + \nu\partial_{yy} u, \tag{5.6}$$

$$\partial_x u + \partial_y v \;=\; 0, \tag{5.7}$$

where $p \equiv p(x) = p^{ext}(x, 0)$. Find the first term on the right-hand side of equation (5.6). From the x-dependence of this term, deduce that every term in equation (5.6) must scale linearly with x.

3. Thus deduce that u scales linearly with x and v is independent of x, i.e. the boundary layer thickness δ is x-independent. Use a dimensional argument to find δ in terms of ν and α.

4. Express v in terms of a non-dimensional function f of non-dimensional variable $\eta = y/\delta$:
$$v = v^* f(\eta).$$
Using a dimensional argument, find v^* in terms of ν and α.

5. Find u in terms of f using the incompressibility condition (5.7).

6. Substitute u and v into the equation (5.6) and find the ordinary differential equation for $f(\eta)$.

7. Use the no-slip boundary conditions on the plate and the conditions of matching to the exterior inviscid flow to find the boundary conditions for $f(\eta)$.

8. You might like to solve the found equation for $f(\eta)$ with the obtained initial conditions using e.g. MATLAB®'s ODE45 programme and thereby

find the profile of the boundary layer. (This could be useful if the problem is offered as a homework in a fluid dynamics course where the use of computers is encouraged).

5.2.4 A flow near an oscillating wall

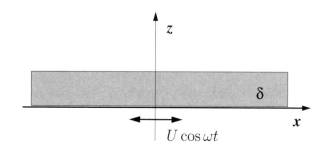

FIGURE 5.4: Boundary layer near an oscillating wall.

An incompressible viscous fluid with viscosity ν occupies the space $0 < y < \infty$ above a plane rigid boundary $y = 0$ which oscillates in the x-direction with velocity $U \cos \omega t$ where U and ω are constants; see figure 5.4.

1. Formulate the no-slip boundary condition for this problem. Formulate the boundary condition at $y \to \infty$.

2. Prove that there is no y-component of velocity in this flow, $v \equiv 0$.

3. Assuming that there is no applied pressure gradient, show that the x-component of the velocity, $u \equiv u(y, t)$, satisfies the equation

$$\partial_t u = \nu \frac{\partial^2 u}{\partial y^2}, \tag{5.8}$$

   where ν is the kinematic viscosity coefficient.

4. Using substitution $u(y, t) = f(y) e^{-i\omega t} + c.c.$ (where *c.c.* denotes complex conjugate), find the solution for the boundary layer velocity.

5. Estimate the boundary layer thickness δ.

5.2.5 A boundary layer in a rotating fluid

Given information: The flow equation in a uniformly rotating frame, equation (1.12).

This problem considers a boundary layer forming in a rapidly rotating

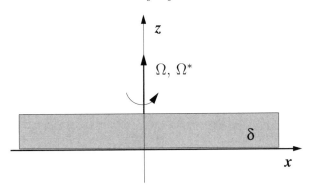

FIGURE 5.5: Boundary layer in a rotating fluid.

fluid—the so-called Ekman boundary layer. Such boundary layers model atmospheric flows near surfaces of rotating planets, e.g. the Earth.

Consider a semi-infinite flow over an infinite plate at $z = 0$ uniformly rotating with angular velocity $\mathbf{\Omega}$ directed normally to the plate, $\mathbf{\Omega} = (0, 0, \Omega)$; see figure 5.5. In the inviscid exterior (i.e. far enough from the plate) the flow itself is rotating with angular velocity $\mathbf{\Omega}^* = (0, 0, \Omega^*)$, which is close to the plane's rotation, $|\Omega^* - \Omega| \ll \Omega$. There is also a constant z-component of velocity in the exterior flow w_{ext} which, as we will see later, will be determined by the rotation angular velocities.

1. Write down an expression for the velocity of the flow uniformly rotating with angular velocity $\mathbf{\Omega}^* = (0, 0, \Omega^*)$.

2. Write an expression for the Rossby number, i.e. the ratio of the typical values of the convective nonlinearity and the Coriolis terms in equation (1.12). What can you say about the value of the Rossby number under the condition $|\Omega^* - \Omega| \ll \Omega$? What term in equation (1.12) can be neglected?

3. Formulate the no-slip boundary condition for this problem. Formulate the boundary condition at $z \to \infty$.

4. Find the pressure field in the exterior inviscid flow, p^{ext}.

5. Use dimensional analysis and find the characteristic thickness of the boundary layer δ assuming that the only relevant dimensional quantities in our problem are Ω and ν.

6. Consider the steady-state flow $\mathbf{u} = (u, v, w)$ in the boundary layer at the distance r from the axis of rotation such that $r \gg \delta$. From the incompressibility condition, deduce that $|w| \ll |u|$ and $|w| \ll |v|$.

7. Use condition $r \gg \delta$ to simplify the viscosity terms in the components

of the momentum equations. Argue that the viscosity term can be neglected in the z-momentum equation and, therefore, pressure p is approximately independent of z and equal to its exterior value p_{ext}.

8. Substitute $p = p_{ext}$ into the x- and y-momentum equations and the
boundary conditions you found before and solve for u and v.

9. Use the incompressibility condition and the solution for u and v and find
w. Using the boundary conditions find the relation for w_{ext} in terms of
Ω, Ω^* and ν.

5.3   Solutions

5.3.1   Model solution to question 5.2.1

1. The viscous term is important in the boundary layer if it is of the same
order as the inertial term:

$$|\nu \nabla^2 \mathbf{u}| \sim |(\mathbf{u} \cdot \nabla)\mathbf{u}|.$$

Considering the fact that the boundary layer flow is nearly plane-parallel
with $u \sim U$ and $v \sim u\delta/L \ll u$ (the latter follows from the incompressibility condition), we have

$$\nu U/\delta^2 \sim U^2/L \sim vU/\delta.$$

Thus

$$\delta \sim L/Re^{1/2}.$$

This relation confirms that $\delta \ll L$ when $Re \gg 1$.

2. The stationary two-dimensional Navier-Stokes equations written in components are

$$u\frac{\partial u}{\partial x} + v\frac{\partial u}{\partial y} = -\frac{\partial p}{\partial x} + \nu\frac{\partial^2 u}{\partial x^2} + \nu\frac{\partial^2 u}{\partial y^2}, \tag{5.9}$$

$$u\frac{\partial v}{\partial x} + v\frac{\partial v}{\partial y} = -\frac{\partial p}{\partial y} + \nu\frac{\partial^2 v}{\partial x^2} + \nu\frac{\partial^2 v}{\partial y^2}, \tag{5.10}$$

$$\frac{\partial u}{\partial x} + \frac{\partial v}{\partial y} = 0. \tag{5.11}$$

In terms of the non-dimensional and scaled variables we have

$$u'\frac{\partial u'}{\partial x'} + v'\frac{\partial u'}{\partial y'} = -\frac{\partial p'}{\partial x'} + \frac{1}{Re}\frac{\partial^2 u'}{\partial x'^2} + \frac{\partial^2 u'}{\partial y'^2},$$

$$\frac{1}{Re^{1/2}}\left[u'\frac{\partial v'}{\partial x'} + v'\frac{\partial v'}{\partial y'}\right] = -Re^{1/2}\frac{\partial p'}{\partial y'} + \frac{1}{Re^{3/2}}\frac{\partial^2 v'}{\partial x'^2} + \frac{1}{Re^{1/2}}\frac{\partial^2 v'}{\partial y'^2},$$

$$\frac{\partial u'}{\partial x'} + \frac{\partial v'}{\partial y'} = 0.$$

From this system in the limit $Re \to \infty$ in the leading order we obtain the reduced system (5.1), (5.2) and (5.3).

3. Integrating equation (5.2) and writing the equations in terms of the original variables we have:

$$u\frac{\partial u}{\partial x} + v\frac{\partial u}{\partial y} = -\frac{\partial p}{\partial x} + \nu\frac{\partial^2 u}{\partial y^2}, \tag{5.12}$$

$$p \equiv p(x) \qquad - \text{ independent of } y, \tag{5.13}$$

$$\frac{\partial u}{\partial x} + \frac{\partial v}{\partial y} = 0. \tag{5.14}$$

4. First of all, we have the no-slip boundary conditions at the solid surface:

$$\mathbf{u} \equiv (u, v) = 0 \quad \text{at} \quad y = 0.$$

The conditions at $y \to \infty$ correspond to matching to the ideal exterior flow. Note that because of the scale separation the matching range can be considered as $y \to \infty$ for the boundary layer solution and as $y \to 0$ for the exterior flow. Thus $\mathbf{u}$ must be matched to the velocity slip of the exterior flow at the boundary, and $p \equiv p(x)$ must be matched to the pressure of the exterior flow at the boundary:

$$[\mathbf{u}]_{y\to\infty} = \left[\mathbf{u}^{ext}\right]_{y\to 0}, \quad p(x) = \left[p^{ext}\right]_{y\to 0}.$$

5.3.2 Model solution to question 5.2.2

1. Because the semi-infinite flat plate does not have a length scale L, the only remaining relevant quantity having the dimension of length is the distance from the plate's edge x. Thus,

$$Re = \frac{Ux}{\nu}, \quad \text{and} \quad \delta = \frac{x}{Re^{1/2}} = \sqrt{\frac{\nu x}{U}}.$$

2. Thus the similarity variable is $\eta = y/\delta = y\sqrt{U/\nu x}$.

The fact that u must match U at $y \to \infty$ suggests that we should be seeking solution in the form

$$u = Uf(\eta)$$

with boundary conditions

$$f(\eta) = 0 \text{ at } \eta = 0, \quad \text{and } f(\eta) = 1 \text{ at } \eta \to \infty, \qquad (5.15)$$

which follow from the no-slip condition at $y = 0$ and matching to the exterior ideal flow respectively. On the other hand, from the incompressibility condition $\partial_x u = -\partial_y v$, we have $v \sim u\delta/x \ll u$ (c.f. problem 5.2.1). This suggests that for v we should take

$$v = \frac{U\delta}{x}g(\eta) = \sqrt{\frac{\nu U}{x}}g(\eta),$$

with boundary conditions

$$g(\eta) = 0 \text{ at } \eta = 0, \quad \text{and } g(\eta) \to 0 \text{ at } \eta \to \infty, \qquad (5.16)$$

which follow from the no-slip condition at $y = 0$ and matching to the exterior ideal flow respectively.

3. From the incompressibility condition we have:

$$\frac{\eta}{2}f'(\eta) = g'(\eta). \qquad (5.17)$$

4. Because the external ideal flow is uniform, it has a constant pressure which is imposed through the boundary layer. Thus the pressure gradient term in the boundary layer equation is zero. Substituting for u and v in terms of f and g into (5.4), we have

$$-\frac{\eta}{2}ff' + gf' = f''. \qquad (5.18)$$

Combining equations (5.17) and (5.18), we have a single equation for the profile f:

$$-\frac{f}{2} = \left(\frac{f''}{f'}\right)', \qquad (5.19)$$

which must be solved with three boundary conditions: two conditions (5.15) and a third one arising from (5.16). To find the latter we write from (5.17)

$$g(\eta) = \frac{1}{2}\int_0^\eta \tilde{\eta}f'(\tilde{\eta})\,d\tilde{\eta}.$$

This automatically gives $g(0) = 0$, whereas the condition $g(\infty) = 0$ implies

$$\int_0^\infty \eta f'(\eta)\,d\eta = 0,$$

which is the required third condition on f.

5.3.3 Model solution to question 5.2.3

1. From Bernoulli's theorem we have

$$p^{ext} = \text{const} - \frac{\rho}{2}(\alpha^2 x^2 + \alpha^2 y^2).$$

2. For the first term on the right-hand side of equation (5.6) we have

$$-\frac{1}{\rho}\partial_x p = \alpha^2 x.$$

The linear x-dependence of this term implies that every term in equation (5.6) must be $\propto x$. This can be realised when $u \propto x$ and v is x-independent.

3. The only combination of ν and α with the dimension of the length is $(\nu/\alpha)^{1/2}$, so

$$\delta \sim \sqrt{\frac{\nu}{\alpha}}.$$

4. The only combination of ν and α with the dimension of the velocity is $(\nu\alpha)^{1/2}$, so

$$v^* \sim \sqrt{\nu\alpha} = \alpha\delta.$$

5. Thus we consider a self-similar solution

$$v = -\alpha\delta f(\eta),$$

where the minus sign is chosen so that $f > 0$.

6. From the expression for v and the incompressibility condition we have

$$\partial_y v = -\partial_x u = -\alpha f'(\eta).$$

Thus

$$u = \alpha x f'(\eta).$$

The integration constant here is set to zero to allow matching to the inviscid exterior (see below).

7. Substituting u, v and the previously found pressure term into equation (5.6), we have

$$\alpha x f' \alpha f' - \alpha\delta f \alpha x \frac{1}{\delta} f'' = \alpha^2 x + \nu \frac{\alpha x}{\delta^2} f''',$$

or

$$(f')^2 - ff'' = 1 + f'''.$$

8. The no-slip boundary conditions are $u(0) = v(0) = 0$ at $y = 0$. From $u(x, 0) = 0$ we have

$$f'(0) = 0.$$

From $v(x, 0) = 0$ we have

$$f(0) = 0.$$

Matching to the external inviscid flow gives

$$\lim_{y/\delta \to \infty} = \alpha x,$$

so

$$f'(\infty) = 1.$$

5.3.4 Model solution to question 5.2.4

1. The no-slip boundary condition for the velocity field $\mathbf{u}(x, y, t)$ is as follows:

$$\mathbf{u}(x, 0, t) = U \cos \omega t. \tag{5.20}$$

The boundary condition at infinity is

$$\mathbf{u}(x, y, t) \to 0 \quad \text{at} \quad y \to \infty. \tag{5.21}$$

2. Our system is invariant with respect to x-shifts, so u is independent of x, and using the incompressibility condition, we get $v \equiv 0$.

3. For the solution see the derivation of the equation (6.14) in section 6.1. In our case there is no pressure gradient, so we have to put $f = 0$ in this equation.

4. Substituting $u(y, t) = f(y) e^{-i\omega t} + c.c.$ into equation (5.8) we get:

$$-i\omega f = \nu f'',$$

where the double prime stands for the second derivative with respect to y. The solution of this equation decaying at $y \to \infty$ is

$$f(y) = C e^{(i-1)\sqrt{\frac{\omega}{2\nu}}y},$$

where C is a constant. Substituting this expression into $u(y, t) = f(y) e^{-i\omega t} + c.c.$ and using the boundary condition (5.21), we get the solution for the boundary layer velocity:

$$u(y, t) = U e^{-\sqrt{\frac{\omega}{2\nu}}y} \cos\left(\sqrt{\frac{\omega}{2\nu}}y - \omega t\right).$$

5. The boundary layer thickness is determined by the decay distance of the exponential function in the above expression:

$$\delta \sim \sqrt{\frac{2\nu}{\omega}}.$$

5.3.5   Model solution to question 5.2.5

1. The velocity of the flow uniformly rotating with angular velocity $\mathbf{\Omega}^* = (0, 0, \Omega^*)$ (measured in the frame rotating with angular velocity $\mathbf{\Omega} = (0, 0, \Omega)$) is

$$\mathbf{u}_{ext\perp} = (\mathbf{\Omega}^* - \mathbf{\Omega}) \times \mathbf{x}. \tag{5.22}$$

2. The Rossby number is considered in question 1.3.3. It is defined by expression (1.16). Using the expression (5.22), we can rewrite expression (1.16) as:

$$Ro = |\Omega^* - \Omega|/|\Omega|. \tag{5.23}$$

Thus, the Rossby number is small, $Ro \ll 1$, if $|\Omega^* - \Omega| \ll \Omega$. Under this condition the nonlinear term $(\mathbf{u} \cdot \nabla)\mathbf{u}$ in equation (1.12) can be neglected:

$$\partial_t \mathbf{u} = -\frac{1}{\rho}\nabla p_R - 2\,\mathbf{\Omega} \times \mathbf{u} + \nu\nabla^2\mathbf{u}, \tag{5.24}$$

where $p_R = p - \Omega^2 r^2/2$.

3. Since equation (1.12) is for the frame rotating with the angular velocity $\mathbf{\Omega}$, the no-slip boundary condition is simply

$$\mathbf{u} = 0 \quad \text{at} \ z = 0. \tag{5.25}$$

The boundary condition at $z \to \infty$ amounts to matching to the inviscid exterior flow:

$$\mathbf{u} \to \mathbf{u}_{ext\perp} + w_{ext}\,\hat{\mathbf{z}} \quad \text{at} \ z \to \infty. \tag{5.26}$$

4. Neglecting the viscosity term in the stationary version of equation (5.24), for the x- and y-components we have:

$$-2\Omega v_{ext} = -2\Omega(\Omega^* - \Omega)x \ = \ -\frac{1}{\rho}\partial_x p_R^{ext}. \tag{5.27}$$

$$2\Omega u_{ext} = -2\Omega(\Omega^* - \Omega)y \ = \ -\frac{1}{\rho}\partial_y p_R^{ext}. \tag{5.28}$$

Integrating these equations we have:

$$p^{ext} = p_R^{ext} + \Omega^2 r^2/2 = p_0 + \Omega(\Omega^* - \Omega/2)r^2 \approx p_0 + \Omega^2 r^2/2,$$

where p_0 is a constant.

5. The physical dimension of Ω is $1/T$ and the physical dimension of ν is L^2/T. The only combination of Ω and ν having the dimension of length is $\sqrt{\nu/\Omega}$, so

$$\delta \sim \sqrt{\nu/\Omega}.$$

6. We have $\partial_x u \sim u/r$, $\partial_y v \sim v/r$ and $\partial_z w \sim w/\delta \gg w/r$. From the incompressibility condition we get $\partial_x u \sim \partial_y v \sim \partial_z w$, so we conclude that $|w| \ll |u|$ and $|w| \ll |v|$.

7. Using the condition $r \gg \delta$, we can estimate

$$\partial_{zz} u \sim u/\delta^2 \gg u/r^2 \sim \partial_{xx} u \sim \partial_{yy} u$$

and

$$\partial_{zz} v \sim v/\delta^2 \gg v/r^2 \sim \partial_{xx} v \sim \partial_{yy} v.$$

Therefore, only z-derivative terms can be retained in the viscosity terms in the momentum equations:

$$-2\Omega v \quad = \quad -\frac{1}{\rho}\partial_x p_R + \nu\partial_{zz} u, \tag{5.29}$$

$$2\Omega u = \quad = \quad -\frac{1}{\rho}\partial_y p_R + \nu\partial_{zz} v, \tag{5.30}$$

$$0 = \quad = \quad -\frac{1}{\rho}\partial_z p_R + \nu\partial_{zz} w. \tag{5.31}$$

Because $w \ll u, v$, the viscosity term in the z-momentum equation is much less than the viscosity terms in the x- and y-momentum equations and, therefore, can be neglected. Therefore, the reduced pressure p_R is approximately independent of z and equal to its exterior value p_R^{ext}.

8. Thus, the pressure terms can be substituted from the exterior equations (5.27) and (5.28):

$$-2\Omega(v - v_{ext}) \quad = \quad \nu\partial_{zz} u, \tag{5.32}$$
$$2\Omega(u - u_{ext}) = \quad = \quad \nu\partial_{zz} v. \tag{5.33}$$

Introducing a complex variable

$$f = u - u_{ext} + i(v - v_{ext}),$$

we have the following complex equation:

$$\nu\partial_{zz} f = 2\Omega i f.$$

Since the velocity must match the exterior velocity value at large z, we have to choose a solution for f which is decaying at $z \to \infty$:

$$f = A e^{-(1+i)z/\delta},$$

where A is an arbitrary function of x and y which has to be fixed by the no-slip boundary condition $u = v = 0$ at $z = 0$, i.e. $f = -u_{ext} - iv_{ext}$ at $z = 0$. Thus

$$A = -u_{ext} - iv_{ext}$$

and for the velocity components we have:

$$u \quad = \quad u_{ext} - e^{-z/\delta}(u_{ext}\cos(z/\delta) + v_{ext}\sin(z/\delta)), \tag{5.34}$$
$$v \quad = \quad v_{ext} - e^{-z/\delta}(v_{ext}\cos(z/\delta) - u_{ext}\sin(z/\delta)). \tag{5.35}$$

9. Using the incompressibility condition we have:

$$w = -\int_0^z (\partial_x u(x, y, z') + \partial_y v(x, y, z')) \, dz'.$$

Substituting here the solution for u and v and integrating to $z = \infty$ using the matching condition $w|_{z=\infty} = w_{ext}$, we get the following relation,

$$w_{ext} = \delta(\partial_x v_{ext} - \partial_y u_{ext})/2 = (\Omega^* - \Omega)\delta.$$

Note that this expression is positive for cyclones, $\Omega^* > \Omega$, and negative for anticyclones, $\Omega^* < \Omega$. This effect is called Ekman pumping.

Chapter 6

Two-dimensional flows

6.1 Background theory

In this chapter, we will consider the most basic two-dimensional (2D) flows in ideal and viscous incompressible fluids. In 2D flows, the velocity field has two components which depend on two physical space coordinates and time, $\mathbf{u} = (u(x, y, t), v(x, y, t), 0)$. Respectively, the vorticity field has only one non-zero component, $\boldsymbol{\omega} = (0, 0, \omega(x, y, t))$. This component satisfies the equation (2.21), which we will reproduce here for convenience:

$$D_t \omega \equiv (\partial_t + (\mathbf{u} \cdot \nabla)) \omega = \nu \nabla^2 \omega. \tag{6.1}$$

For 2D incompressible flows, one can introduce a representation of the velocity field in terms of stream function $\psi(x, y, t)$ as follows,

$$\mathbf{u} = \nabla \psi \times \hat{\mathbf{z}}, \tag{6.2}$$

or in component form

$$u = \partial_y \psi, \quad v = -\partial_x \psi. \tag{6.3}$$

In terms of the stream function the vorticity is:

$$\omega = -\nabla^2 \psi. \tag{6.4}$$

If the 2D flow is irrotational, then we also have a representation in terms of the velocity potential $\mathbf{u} = \nabla \phi$, or

$$u = \partial_x \phi, \quad v = \partial_y \phi, \tag{6.5}$$

and the viscous term is automatically zero, $\nu \nabla^2 \mathbf{u} = \nu \nabla \nabla^2 \phi = 0$.

Combining expressions (6.3) and (6.5) we have

$$\partial_x \phi = \partial_y \psi, \quad \partial_y \phi = -\partial_x \psi. \tag{6.6}$$

These relations are nothing but the Cauchy-Riemann conditions, i.e. necessary and sufficient conditions for a complex function $w(z) = \phi + i\psi$ (where $z = x + iy$) to be complex differentiable, that is, holomorphic. The function $w(z)$

is called the complex potential. Both real and imaginary parts of it satisfy Laplace's equation,

$$\nabla \cdot \mathbf{u} = \nabla^2 \phi = 0, \quad \nabla \times \mathbf{u} = \nabla^2 \psi = 0. \tag{6.7}$$

For the velocity $\mathbf{u} = (u, v)$ in terms of the complex potential we have

$$u - iv = \partial_z w, \tag{6.8}$$

so that

$$|\mathbf{u}| = \sqrt{u^2 + v^2} = |\partial_z w|. \tag{6.9}$$

The identification of the complex potential of a 2D ideal irrotational flow with a holomorphic function had profound consequences for the rapid development of both aerodynamic theory and complex analysis, notably utilising conformal maps to obtain new fluid flow solutions out of already known and often simpler ones. This approach culminated in 1906 in Zhukovskiy's lift theorem, stating the following.

A steady ideal flow past a 2D body and having a uniform velocity $U\hat{\mathbf{x}}$ far from the body (i) produces no force component along $\hat{\mathbf{x}}$, i.e. $F_x = 0$, and (ii) produces a y-component of the force (per unit length along in the transverse to the 2D plane direction, i.e. $\hat{\mathbf{z}}$) equal to

$$F_y = -\rho U \Gamma, \tag{6.10}$$

where Γ is the velocity circulation around a contour embracing the 2D body (e.g. the 2D body's boundary). Part (i) asserts the absence of a drag force and is often referred to as D'Alembert's paradox. Part (ii) is a statement about the lift force.

A proof of this theorem, as well as a detailed discussion of the complex analysis application to aerodynamics and the aerofoil theory can be found in Acheson's book [1].

Let us now consider viscous flows. The most basic 2D flows have plane-parallel configuration in which the velocity field is everywhere in the same direction and constant along that direction, i.e.

$$\mathbf{u}(\mathbf{x}, t) = (u(y, t), 0, 0). \tag{6.11}$$

The incompressibility condition for such a velocity field is automatically satisfied, $\nabla \cdot \mathbf{u} = 0$, and the nonlinear term $(\mathbf{u} \cdot \nabla)\mathbf{u}$ is automatically zero, which greatly simplifies finding solutions. Writing the Navier-Stokes equation in components, we have

$$\partial_t u = -\frac{1}{\rho}\partial_x p + \nu \partial_{yy} u, \tag{6.12}$$

$$0 = -\frac{1}{\rho}\partial_y p. \tag{6.13}$$

From the second equation we see that the pressure is independent of y and from the first one, that its dependence on x is linear, i.e. the pressure force is uniform and in the x-direction, $-\nabla p/\rho = f\hat{\mathbf{x}}$ where $f = \text{const}$. Thus we have the following equation

$$\partial_t u = f + \nu\partial_{yy}u. \tag{6.14}$$

Introducing a new variable

$$\tilde{u} = u - \frac{fy^2}{2\nu} \tag{6.15}$$

we get the 1D linear diffusion equation

$$\partial_t\tilde{u} = \nu\partial_{yy}\tilde{u}, \tag{6.16}$$

which is a classical and a very well studied equation with many solution-finding techniques developed for it.

One can also rewrite equation (6.14) as a continuity equation corresponding to the x-momentum conservation:

$$\partial_t(\rho u) + \nabla \cdot \mathbf{F} = 0, \tag{6.17}$$

where $\mathbf{F}$ is the x-momentum flux:

$$\mathbf{F} = -f\rho x\,\hat{\mathbf{x}} - \nu\rho(\partial_y u)\,\hat{\mathbf{y}}. \tag{6.18}$$

The first part of the flux is due to the pressure forces and the second part corresponds to the x-momentum transfer between the moving fluid layers due to the internal friction. As such, the latter part determines the wall friction per unit area, i.e. the drag:

$$D = \nu\rho|\partial_y u|_{boundary}. \tag{6.19}$$

Similarly, one can consider velocity profiles having only an azimuthal component independent of the polar angle, which represent various round vortex solutions. Here, too, the simplification comes from the fact that the nonlinear term is automatically zero.

In the problems below, we will start with the most basic 2D flows, such as the constant-strain and the plane-parallel flows, as well as round vortices. This will be followed by considering some irrotational 2D flows given by a prescribed complex potential. We will also consider some applications of Zhukovskiy's theorem. We will consider more general theoretical questions about the behaviour of vorticity and stream function in 2D flows. Finally, we will consider slightly more complicated flows down inclined slopes.

6.2   Problems

6.2.1   Pure strain flow

A two-dimensional flow field is described by the velocity components $u = \alpha x$ and $v = -\alpha y$, where α is a positive constant.

1. What is the vorticity field of this flow?

2. Compute the velocity potential, the stream function and the complex potential for the flow.

3. Find an expression for the streamlines of the flow.

4. A particle of dust is placed at time $t_0 = 0$ at the point (x_0, y_0) on an arbitrary streamline. At what time t_1 does the dust particle reach the point (x_1, y_1) of the streamline? It is assumed that the dust particle has a very small mass, so that no slippage occurs between it and the flow.

6.2.2   Couette flow

Consider a plane shear flow, $\mathbf{u} = (u(y), 0, 0)$, between two infinite plates at $y = h$ and $y = -h$, which are moving in the x-direction with velocities U and $-U$ respectively; see figure 6.1. Pressure in the flow is uniform in space (the flow is not pressure driven).

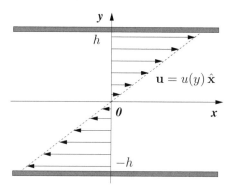

FIGURE 6.1: Couette flow.

1. Show that any profile $u(y)$ gives a solution to the Euler equation with the free-slip boundary conditions.

2. Take viscosity into account. State the no-slip boundary conditions that must be satisfied by solutions of the Navier-Stokes equations at the plates.

3. Find the steady-state velocity profile $u(y)$ that satisfies the Navier-Stokes equations with the no-slip boundary conditions at the moving plates.

4. Find the friction force per unit area produced by the flow on the plates, expressed in terms of U, h and density ρ and viscosity coefficient ν.

6.2.3   Poiseuille flow

This problem is similar to the previous one, but the flow is now pressure driven and it occurs between fixed boundaries.

Consider a plane-parallel flow, $\mathbf{u} = (u(y), 0, 0)$, between two infinite plates at $y = h$ and $y = -h$ which are fixed (not moving); see figure 6.2. The flow is driven by a uniform pressure gradient in the x-direction, $-\nabla p/\rho = f\hat{\mathbf{x}}$ where $f = \text{const}$.

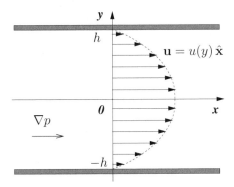

FIGURE 6.2: Poiseuille flow.

1. Find the steady-state velocity profile $u(y)$ that satisfies the Navier-Stokes equations with the no-slip boundary conditions at the fixed plates.

2. Find the friction force per unit area produced by the flow on the plates, expressed in terms of f, h and density ρ and viscosity coefficient ν.

6.2.4   "Turbulent" shear flows

Given information: the Navier-Stokes equation for a medium with non-uniform viscosity coefficient $\nu(\mathbf{x})$ is

$$\partial_t \mathbf{u} + (\mathbf{u} \cdot \nabla)\mathbf{u} = -\frac{1}{\rho}\nabla p + \nabla \left(\nu(\mathbf{x})\nabla \mathbf{u}\right). \qquad (6.20)$$

In questions 6.2.2 and 6.2.3 we considered laminar plane-parallel shear flows between two infinite parallel plates: Couette and Poiseuille flows. At very high Reynolds numbers, such flows become unstable and the instability leads to turbulence. At a very basic level, turbulence effects upon the mean velocity profile may be thought of as an enhanced viscosity—the so-called eddy viscosity. In this problem we will study such an effect in Couette and Poiseuille flows by postulating a non-uniform viscosity profile so that the latter is greater in the central part of the flow, which is most turbulent:

$$\nu \equiv \nu(y) = \lambda - \mu y^2, \tag{6.21}$$

with $\lambda > \mu h^2$, so that ν is positive at the walls $y = \pm h$ (where there is no turbulence and ν is the same as in the laminar flow).

1. Consider a plane-parallel shear flow, $\mathbf{u} = (u(y), 0, 0)$, between two infinite plates at $y = h$ and $y = -h$, which are moving in the x-direction with velocities U and $-U$, respectively as in figure 6.1. Pressure in the flow is uniform in space (the flow is not pressure driven).

   Find the steady-state velocity profile $u(y)$ that satisfies the Navier-Stokes equations with the no-slip boundary conditions at the moving plates.

2. Find the friction force per unit area produced by the flow on the plates. Is it greater or less than the friction produced by the laminar Couette flow with constant viscosity corresponding the viscosity value in the laminar part of the flow (i.e. near the walls) $\nu = \lambda - \mu h^2$?

3. Now consider a plane-parallel flow, $\mathbf{u} = (u(y), 0, 0)$, between two infinite plates at $y = h$ and $y = -h$ which are not moving; see figure 6.2. The flow is driven by a uniform pressure gradient in the x-direction, $-\nabla p / \rho = f \hat{\mathbf{x}}$ where $f = \text{const}$.

   Find the steady-state velocity profile $u(y)$ that satisfies the Navier-Stokes equations with the no-slip boundary conditions at the fixed plates.

4. Find the friction force per unit area produced by the flow on the plates. Is it greater or less than the friction produced by the laminar Poiseuille flow with $\nu = \lambda - \mu h^2$?

5. Sketch the mean flow profiles you have obtained for the turbulent Couette and Poiseuille flows. Comment on the qualitative differences of the mean velocity profiles of the considered flows in the laminar and the turbulent states near the walls and in the centre of the channel. Is the shear reduced or increased at these locations?

6.2.5 Jet flow

Consider a plane-parallel flow, $\mathbf{u} = (u(y,t),0,0)$, in an infinite unbounded 2D space, whose initial profile is a narrow jet; see figure 6.3. Namely, at $t = 0$ it has non-zero velocity only in a small vicinity of $y = 0$ and it could be approximated by a delta function:

$$u(y) = A\,\delta(y),$$

where A is a real constant. As time goes on, the jet profile is spreading due to the action of the viscosity, and it can be described by a self-similar solution to equation (6.14) in which $f = 0$ (there is no pressure gradient).

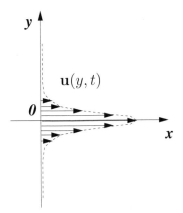

FIGURE 6.3: Jet flow.

1. Find the physical dimensions of the kinematic viscosity coefficient ν and of the constant A.

2. Explain why the similarity variable is not dependent on A but the solution $u(y,t)$ is. What kind of dependence on A is it?

3. Find the similarity variable η.

4. Find the self-similar solution for $u(y,t)$.

6.2.6 Mixing layer

This problem is similar to the previous one in the way it is solved, but one would have to satisfy different boundary conditions. In fact, the boundary conditions for the vorticity field would be the same as the ones for the velocity field in the previous problem.

The present question is also very similar to the problem of the flow due to an infinite plane which suddenly sets in a constant velocity motion.

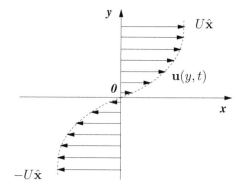

FIGURE 6.4: Mixing layer.

Consider a plane-parallel flow, $\mathbf{u} = (u(y,t), 0, 0)$, in an infinite unbounded 2D space, whose initial profile has a tangential discontinuity. Namely, at $t = 0$ it has velocity $u(y) = U$ at $y > 0$ and $u(y) = -U$ at $y < 0$ where U is a real constant. At $t > 0$ the discontinuity is replaced by a layer where the x-velocity gradually transitions from $-U$ to U; see figure 6.4. It is called a mixing layer, and it is expanding in time due to the action of the viscosity. Its evolution can be described by a self-similar solution to equation (6.14) in which $f = 0$ (the is no pressure gradient).

1. Find the physical dimensions of the kinematic viscosity coefficient ν and of the velocity U.

2. Explain why the similarity variable is not dependent on U but the solution $u(y,t)$ is. What kind of dependence on U is it?

3. Find the similarity variable η.

4. Find the self-similar solution for $u(y,t)$.

6.2.7   Stream function for a 2D flow

A stationary 2D incompressible (viscous or inviscid) flow is described by a stream function $\psi(x,y)$ such that the velocity is

$$\mathbf{u} = (u(x,y), v(x,y), 0) = \left(\frac{\partial\psi}{\partial y}, -\frac{\partial\psi}{\partial x}, 0\right).$$

Show that:

1. The streamlines are given by $\psi = \text{const}$.

2. $|\mathbf{u}| = |\nabla\psi|$, and conclude that the flow is faster where the streamlines are closer.

3. The volume flux (per unit length in z) crossing any curve from (x_1, y_1) to (x_2, y_2) is given by $\psi(x_2, y_2) - \psi(x_1, y_1)$.

4. $\psi = $ const on any fixed (i.e. stationary) boundary.

5. Find the stream function for the flow with the following velocity field,

$$u = \frac{y}{x^2 + y^2}, \quad v = -\frac{x}{x^2 + y^2}.$$

Sketch the streamlines and describe the properties of this flow.

6.2.8 Round vortices: Rankine vortex and a point vortex

This problem deals with round 2D vortices, i.e. vortices whose vorticity distributions depend only on the distance from the vortex centre and not on the polar angle.

1. Consider a vorticity field of a 2D flow, $\boldsymbol{\omega} = (0, 0, \omega)$, such that in polar coordinates ω depend only on the radius and not on the polar angle, $\omega \equiv \omega(r)$ where $r = \sqrt{x^2 + y^2}$.

   Prove that such a vorticity distribution represents a steady-state solution of the ideal flow equations for any profile $\omega(r)$. (This may not necessarily be a stable, i.e. a realisable solution).

2. *Rankine vortex.* Consider a vortex profile $\omega(r)$ such that $\omega(r) = \kappa = $ const for $r \leq a$ ($a = $ const) and $\omega(r) = 0$ for $r > a$. Find the velocity and the pressure fields for such a vortex.

3. Take the limit $a \to 0$ while keeping the combination $\Gamma = \kappa a^2$ constant (vortex circulation). This limit corresponds to the so-called point vortex. Find the velocity and the pressure fields for such a vortex.

6.2.9 Flow bounded by two intersecting planes

Consider a 2D potential flow described by the complex potential

$$w(z) = Cz^n,$$

where C and n are real positive numbers.

1. Suppose that $n = 1$. Describe the resulting flow and the meaning of the constant C.

2. Suppose now that $n = 2$. Describe the resulting flow and the meaning of the constant C.

3. Now consider the case where n is an arbitrary rational number which is greater or equal to $1/2$. Write down expressions for the velocity potential ϕ and the stream function ψ in terms of the polar coordinates r and θ on the complex plane, $z = re^{i\theta}$. Find the smallest and the second-smallest positive angles, $\theta = \theta_1$ and $\theta = \theta_2$, for which the stream function ψ is zero for any r. Consider the flow in the sector $\theta_1 < \theta < \theta_2$, i.e. between the two intersecting (at $z = 0$) planes $\theta = \theta_1$ and $\theta = \theta_2$. Explain why the flow with the complex potential $w(z) = Cz^n$ satisfies the free-slip boundary conditions at these planes.

4. Sketch the flows arising in the cases $n = 4$, $n = 4/3$, $n = 2/3$ and $n = 1/2$.

5. Describe the qualitative difference in behaviour of the velocity field near $r = 0$ for cases $n < 1$ and $n > 1$.

6.2.10 "Binary star system"

The setup of this problem could be considered a simplified 2D analogue of the flow arising in a binary star system in which one of the stars is a rotating black hole. The reader is recommended to search the web for Cygnus X-1 images for illustration.

1. Consider a 2D potential flow described by the complex potential

$$w(z) = C \ln z,$$

where C is a complex constant. Describe the flows when C is real and when C is purely imaginary.

2. Consider a 2D potential flow described by the complex potential

$$w(z) = \ln z - (1 + ia) \ln(z - 1),$$

where a is a real positive constant.

Find the stagnation point in this flow.

3. Describe how this flow behaves at $r = |z| \to \infty$.

4. Sketch the streamlines and describe the general features of the flow.

5. Assume now that the "black hole is not rotating", $a = 0$. Find the shortest lifetime of the fluid particles, i.e. the shortest time between the emergence of a fluid particle at the source at $(x, y) = (0, 0)$ and its disappearance into the "black hole" at $(x, y) = (1, 0)$.

6.2.11 Complex potential for the gravity water waves

The velocity potential corresponding to the gravity wave on deep water is (c.f. question 4.3.2):

$$\phi = Ce^{ky}\cos(kx - \omega t), \quad y < 0, \tag{6.22}$$

where the water surface is at $y = 0$, and C, k and ω are real positive constants having the meaning of the wave amplitude, wave number and frequency.

1. Find the velocity field under the water surface.

2. Find the stream function and the complex potential.

3. Find the trajectories of the fluid particles and the streamlines. Comment on the differences of these two types of curves.

6.2.12 Aeroplane lift and trailing vortices

This problem deals with the lift force on a flying aeroplane, its relation to the velocity circulation around the aerofoil and the trailing vortices originating at the airplane wingtips.

1. Explain why the vortex lines near the wing surface (responsible for the circulation round the aerofoil) must continue into a trailing vortex through the wingtip. What is the relation between the circulation around the wing and circulation round its trailing vortex?

2. An aeroplane of mass M is flying at speed V through air of density ρ. Each wing of this plane is L meters long. Find the circulation around each of its trailing vortices. (**Hint:** use Zhukovskiy's lift theorem.)

3. Assume that the trailing vortices originating at both wings are parallel to each other and that their vorticity is concentrated in thin tubes separated by a distance $D = 2L$. Find the downdraft velocity produced by these trailing vortices at the midpoint between them.

6.2.13 Finding drag and lift using dimensional analysis

Consider a 2D incompressible flow with incident velocity U around a solid body with typical length scale L.

1. Find the physical dimensional unit for the drag and lift forces (per unit length in the direction transverse to the motion plane).

2. Use the dimensional analysis to find the lift force assuming that the relevant dimensional quantities determining this quantity are U, ρ and L (this will be famous Kutta's condition).

Find the lift force in terms of U, ρ and Γ (this will be famous Zhukovskiy's force).

3. Find the drag force assuming that the relevant dimensional quantities determining this quantity are U, ρ and L. This regime may be realised for some blunt bodies with separated flow and turbulence in its wake.

4. What is the most general expression for the drag force assuming that the relevant dimensional quantities determining this quantity are U, ρ, ν and L? (**Hint:** you may use the result of the previous question and use Re instead of ν.) Note: the drag dependence on Re will be $Re^{-1/2}$ for a flat plate. What other dependencies are possible? (**Hint:** think of d'Alembert's paradox.)

6.2.14   A laminar jet flow

A 2D jet emerges from a narrow slit in a wall into fluid which is at rest. If the jet is thin, so that velocity $\mathbf{u} = (u, v)$ varies much more rapidly across the jet than along it, the fluid equation becomes

$$u\frac{\partial u}{\partial x} + v\frac{\partial u}{\partial y} = \nu\frac{\partial^2 u}{\partial y^2}, \qquad (6.23)$$

where constant ν is the viscosity coefficient. The boundary conditions are that the velocity and its derivatives tend to zero as we leave the jet (that is as $|y| \to \infty$) and $\partial u/\partial y = 0$ at $y = 0$, as the motion is symmetrical about the x-axis.

1. By integrating equation (6.23) across the jet and using the incompressibility condition, show that the integral

$$M = \int_{-\infty}^{\infty} u^2\, dy \qquad (6.24)$$

is independent of x.

2. Assuming that the stream function is of a self-similar form,

$$\psi = x^a\, f(\eta), \quad \text{where } \eta = yx^b, \qquad (6.25)$$

find the relation between the constants a and b using part (1).

3. Substitute the similarity solution (6.25) into equation (6.23) and find the second relation between a and b. Thus, find a and b.

4. Show that jet equation (6.23) and the velocity boundary conditions lead to the following equation for $f(\eta)$,

$$(f')^2 + ff' + 3\nu\, f''' = 0$$

subject to the conditions

$$f(0) = f''(0) = 0, \quad f'(\infty) = 0,$$

where prime means differentiation. (It is possible to find an exact solution of this equation but you do not have to do it here.)

6.2.15 Flow in a cylinder with an elliptical cross-section

A cylinder filled with water has an elliptical cross-section with major and minor semi-axes a and b. While $t < 0$ both the cylinder and the water within it rotate about the axis of the cylinder with constant angular velocity Ω. At time $t = 0$ the cylinder is suddenly brought to rest. In this problem you will need to find the water flow that is established inside the cylinder at $t > 0$.

1. For a 2D flow $(u(x,y,t), v(x,y,t), 0)$ find the expression for the vorticity $\boldsymbol{\omega}$ in terms of the stream function ψ.

2. Find the velocity, the stream function and the vorticity of the flow uniformly rotating with angular velocity $\boldsymbol{\Omega}$. (Take $\boldsymbol{\Omega}$ to be along the z-axis.)

3. Write down the evolution equation for vorticity in a 2D ideal flow, and explain why it describes the conservation of vorticity along trajectory of each fluid particle.

4. Use the results of parts (2) and (3) and find the vorticity of the water flow that is set up inside the cylinder at $t > 0$.

5. Use the results of parts (1), (2) and (4), as well as the condition $\psi =$const at the boundary, to find the stream function of the water flow inside the cylinder at $t > 0$. Sketch the streamlines.

6.2.16 Rain flow over an inclined roof

A violent rainstorm hits a roof inclined at an angle θ from the horizontal as shown in figure 6.5. The rain pours down at a mass flow rate Q per unit horizontal area, and each drop falls at a velocity V. Soon a steady-state water (density ρ) layer is established, while raindrops splash on the top part of the thin layer. The angle of the water surface relative to the roof is small $(dh/dx \ll 1)$, and friction between the roof and water may be neglected.

FIGURE 6.5: Flow down a roof.

1. What is the pressure distribution in the water in the direction perpendicular to the roof? Parallel to the roof? Find the pressure within the fluid layer.

2. By considering the balance of the momentum and mass in a small volume of fluid between x and $x + dx$, find the evolution equations for the layer thickness $h(x)$ and the flow velocity $u(x)$.

3. Derive a criterion for when the rain is so violent that the shape of the water layer on the roof can be regarded as independent of gravity. Obtain a solution for $h(x)$ in this case.

6.2.17   Flow over an inclined plane

A layer of viscous fluid is bounded from below by a plane fixed at angle θ to the horizon, and from above by a free surface. The thickness of the layer is h. The air above the free surface is at atmospheric pressure p_0. Consider a stationary flow resulting from the balance between gravity and viscosity; see figure 6.6.

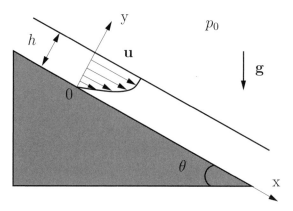

FIGURE 6.6: Flow down a slope.

1. Formulate the no-slip boundary condition for the velocity field at the bottom plane and the pressure boundary condition at the free surface.

2. Show that absence of stress at the free surface means that $\frac{\partial u}{\partial y} = 0$ at $y = h$.

3. Solve the stationary Navier-Stokes equations and find the velocity and the pressure fields (the velocity is parallel to the plane).

4. Find the mass flux per unit length in the transverse to the plane of motion direction, i.e. mass of fluid Q passing through the cross-section of the layer per unit z and per unit time.

6.3 Solutions

6.3.1 Model solution to question 6.2.1

1. The vorticity in 2D flows has only a z-component, $\Omega = \partial_x v - \partial_y u$. Substituting u and v, we have $\Omega \equiv 0$.

2. From $\mathbf{u} = \nabla \phi$, we have for the velocity potential $\phi = \frac{1}{2}\alpha(x^2 - y^2)$. From $\mathbf{u} = (\partial_y \psi, -\partial_y \psi)$, we have the steamfunction $\psi = \alpha xy$. From $w = \phi + i\psi$, we have the complex potential $w(z) = \frac{1}{2}z^2$.

3. Equation for the streamlines:
$$\frac{dx}{u} = \frac{dy}{v}.$$

Substituting u and v:
$$\frac{dx}{x} = -\frac{dy}{y}.$$

Solving for x and y, we get $y = C/x$, where C is an arbitrary constant (parametrising the streamlines).

4.
$$\frac{dx}{dt} = \alpha x,$$

so
$$x = x_0 e^{\alpha t}.$$

Therefore
$$t_1 = \frac{1}{\alpha}\ln\frac{x_1}{x_0}.$$

6.3.2 Model solution to question 6.2.2

1. Let us start with the Euler equation:
$$\partial_t \mathbf{u} + (\mathbf{u} \cdot \nabla)\mathbf{u} = -\frac{1}{\rho}\nabla p.$$

Consider a steady plane-parallel shear flow: $\mathbf{u} = (u(y), 0, 0)$. The flow is time independent,
$$\partial_t \mathbf{u} = 0,$$

and x-independent,
$$(\mathbf{u} \cdot \nabla)\mathbf{u} = u\partial_x\mathbf{u} = 0.$$

Therefore, the x-component of the Euler equation is satisfied if $\partial_x p = 0$, i.e. p is independent of x.

For the y-component of the Euler equation we have: $0 = -\frac{1}{\rho}\partial_y p$, i.e. p is y-independent. Same for the z-component of the Euler equation: $0 = -\frac{1}{\rho}\partial_z p$, i.e. p is z-independent.

Therefore $\mathbf{u} = (u(y), 0, 0)$ is a solution for any profile $u(y)$ and $p = \text{const}$.

2. The no-slip boundary conditions are: $u(h) = U, \quad u(-h) = -U$.

3. Let us write the Navier-Stokes equation:

$$\partial_t \mathbf{u} + (\mathbf{u} \cdot \nabla)\mathbf{u} = -\frac{1}{\rho}\nabla p + \nu \nabla^2 \mathbf{u}.$$

The left-hand side of this equation and ∇p are zero as before on $\mathbf{u} = (u(y), 0, 0)$. Therefore we have:

$$\nabla^2 u = \partial_{yy} u(y) = 0.$$

The solution of this equation is

$$u(y) = Ay + B.$$

The boundary conditions give $B = 0$ and $A = U/h$, i.e.

$$u(y) = \frac{U}{h}y.$$

4. The friction per unit area at the bottom plate is equal to the x-momentum flux in the y-direction σ_{xy}:

$$\sigma_{xy} = \nu\rho[\partial_y u]_{y=-h} = \frac{U\nu\rho}{h}.$$

The friction at the top plate is equal in strength and opposite in sign.

6.3.3 Model solution to question 6.2.3

1. The right-hand side of the Navier-Stokes equation is zero on any plane-parallel flow $\mathbf{u} = (u(y), 0, 0)$, i.e. $\partial_t \mathbf{u} + (\mathbf{u} \cdot \nabla)\mathbf{u} = 0$; see solution 6.3.2.

Therefore

$$\partial_x p/\rho = -f = \nu\partial_{yy} u(y).$$

The solution of this equation is

$$u(y) = -\frac{f}{2\nu}y^2 + By + C.$$

The no-slip boundary conditions to be satisfied are $u(h) = u(-h) = 0$. Thus $B = 0$ and $C = \frac{f}{2\nu}h^2$, i.e.

$$u(y) = \frac{f}{2\nu}(h^2 - y^2).$$

2. The friction per unit area at the bottom plate is equal to the x-momentum flux in the y-direction σ_{xy}:

$$\sigma_{xy} = [\nu\rho\partial_y u]_{y=-h} = \rho f h.$$

The friction at the top plate is the same.

6.3.4 Model solution to question 6.2.4

1. For the plane-parallel shear flow, $\mathbf{u} = (u(y), 0, 0)$, which has a uniform pressure, $\nabla p = 0$, the Navier-Stokes equation (6.20) becomes:

$$\partial_y(\nu\partial_y u(y)) = \partial_y((\lambda - \mu y^2)\partial_y u) = 0. \tag{6.26}$$

Integrating this equation once, we have:

$$(\lambda - \mu y^2)\partial_y u = A,$$

where A is a constant. Solving this equation, we have:

$$u(y) = \int \frac{A}{\lambda - \mu y^2} \, dy = \frac{A}{\sqrt{\lambda\mu}} \operatorname{arctanh} \frac{\sqrt{\mu}y}{\sqrt{\lambda}} + B, \tag{6.27}$$

where B is a constant. Matching to the no-slip conditions $u(h) = U$ and $u(-h) = -U$ we have $B = 0$ and $A = U\sqrt{\lambda\mu}/\operatorname{arctanh}\frac{\sqrt{\mu}h}{\sqrt{\lambda}}$. Thus,

$$u(y) = \frac{U}{\operatorname{arctanh}\frac{\sqrt{\mu}h}{\sqrt{\lambda}}} \operatorname{arctanh}\frac{\sqrt{\mu}y}{\sqrt{\lambda}}. \tag{6.28}$$

2. The friction per unit area at the bottom plate is equal to the x-momentum flux in the y-direction σ_{xy} (it is convenient here to use expression (6.27)):

$$\sigma_{xy} = [\nu\rho\partial_y u]_{y=-h} = \rho A = \frac{U\rho\sqrt{\lambda\mu}}{\operatorname{arctanh}\frac{\sqrt{\mu}h}{\sqrt{\lambda}}}.$$

The friction force at the top plate has the same absolute value and the opposite sign.

This friction force is greater (in the absolute value) than the friction $U\rho(\lambda - \mu h^2)/h$ produced by the laminar Couette flow with $\nu = \lambda - \mu h^2$ because $\lambda - \mu h^2 < \lambda$ and $\operatorname{arctanh}\frac{\sqrt{\mu}h}{\sqrt{\lambda}} < \frac{\sqrt{\mu}h}{\sqrt{\lambda}}$.

3. Now consider a plane-parallel flow between two parallel plates, which is driven by a uniform pressure gradient, $-\nabla p/\rho = f\hat{\mathbf{x}}$, where $f = \text{const}$. The Navier-Stokes equation (6.20) becomes:

$$\partial_y(\nu\partial_y u(y)) = \partial_y((\lambda - \mu y^2)\partial_y u) = -f. \tag{6.29}$$

Integrating this equation once, we have:

$$(\lambda - \mu y^2)\partial_y u = -fy + A, \tag{6.30}$$

where A is a constant. Solving this equation, we have:

$$u(y) = \int \frac{A - fy}{\lambda - \mu y^2} \, dy = \frac{A}{\sqrt{\lambda\mu}} \text{arctanh} \frac{\sqrt{\mu}y}{\sqrt{\lambda}} + f \ln \frac{\lambda - \mu y^2}{2\mu} + B, \tag{6.31}$$

where B is a constant. Matching to the no-slip conditions $u(h) = u(-h) = 0$ we have $A = 0$ and

$$B = -f \ln \frac{\lambda - \mu h^2}{2\mu},$$

so

$$u(y) = f \ln \frac{\lambda - \mu y^2}{\lambda - \mu h^2}. \tag{6.32}$$

4. For the friction per unit area at the bottom plate we have (it is convenient here to use expression (6.30)):

$$\sigma_{xy} = [\nu\rho\partial_y u]_{y=-h} = \rho f h.$$

This is the same as the friction produced by the laminar Poiseuille flow with $\nu = \lambda - \mu h^2$ (note that both expressions are viscosity independent).

5. Sketches of the mean profiles for the turbulent Couette and Poiseuille flows are shown in figures 6.7 and 6.8. The mean velocity profiles of the considered flows in the laminar states have smaller shear near the walls and larger shear in the middle of the flow than in the turbulent states.

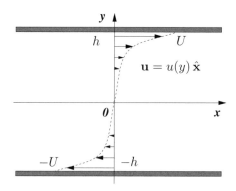

FIGURE 6.7: "Turbulent" Couette flow.

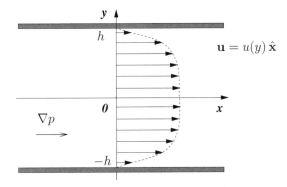

FIGURE 6.8: "Turbulent" Poiseuille flow.

6.3.5 Model solution to question 6.2.5

1. From the formula $u(y) = A\,\delta(y)$ we get the physical dimension of A as L^2/T. The physical dimension of the kinematic viscosity coefficient ν is the same, L^2/T.

2. Because the governing equation (6.14) (with $f = 0$) is linear, the solution $u(y, t)$ is linearly dependent of A. But then the similarity variable is independent of A—otherwise $u(y, t)$ could only be linear in A if it is linear in η, which is clearly not the case.

3. Thus the similarity variable could only depend on the single remaining dimensional quantity—the kinematic viscosity coefficient ν. The only dimensionless combination of ν, y and t is

$$\eta = \frac{y^2}{\nu t}.$$

4. The only combination of A, ν and t which has the velocity dimension and which is linear in A is $A/\sqrt{\nu t}$. Thus we seek a self-similar solution in the form:

$$u(y, t) = \frac{A}{\sqrt{\nu t}}\,F(\eta).$$

Substituting this expression into equation (6.14) (with $f = 0$), we get the following ordinary differential equation for F,

$$-F/2 - \eta F' = 2F' + 4\eta F'',$$

solving which we have

$$F = Ce^{-\eta/4}.$$

The constant C is to be found from matching to the initial condition, i.e. from the relation

$$\int_{-\infty}^{+\infty} u(y, t)\,dy = A.$$

Here we have used the fact that the integral $\int_{-\infty}^{+\infty} u(y, t)\, dy$ is conserved in time.

The easiest way to use this condition is to consider time $t = 1/\nu$ so that $\nu t = 1$. Then we have

$$C \int_{-\infty}^{+\infty} e^{-y^2/4}\, dy = 1,$$

or $C = \frac{1}{2\sqrt{\pi}}$. Finally, we have

$$u(y, t) = \frac{A}{2\sqrt{\pi \nu t}}\, e^{\frac{y^2}{4\nu t}}.$$

6.3.6   Model solution to question 6.2.6

This problem is similar to the previous one in the way it is solved, but one would have to satisfy different boundary conditions. The boundary conditions for the vorticity field will be the same as the ones for the velocity field in the previous problem.

1. The physical dimension $[\nu]$ of the kinematic viscosity coefficient ν can be found from comparing the viscous term in the Navier-Stokes equation, $\nu \nabla^2 \mathbf{u}$, with the time derivative term $\partial_t \mathbf{u}$ (obviously they must have the same physical dimension). Thus $[\nu] = L^2/T$. The physical dimension of the velocity U is L/T.

2. Because the governing equation (6.14) (with $f = 0$) is linear, the solution $u(y, t)$ is linearly dependent on U. But then the similarity variable is independent of U—otherwise $u(y, t)$ could only be linear in U if it is linear in η, which is clearly not the case.

3. Thus the similarity variable could only depend on the single remaining dimensional quantity—the kinematic viscosity coefficient ν. The only dimensionless combination of ν, y and t is

$$\eta = \frac{y^2}{\nu t}.$$

4. We seek a self-similar solution in the form:

$$u(y, t) = U\, F(\eta).$$

Substituting this expression into equation (6.14) (with $f = 0$), we get the following ordinary differential equation for F,

$$-\eta F' = 2F' + 4\eta F''.$$

5. The above equation can be rearranged as:

$$F''/F' = -\frac{1}{2\eta} - \frac{1}{4}.$$

Subsequently integrating, we have:

$$F' = C\eta^{-1/2}e^{-\frac{\eta}{4}},$$

where C is a constant. Integrating one more time, we have:

$$F = C \int \eta^{-1/2}e^{-\frac{\eta}{4}}\,d\eta = 4C \int e^{-\xi^2}\,d\xi = 2C\sqrt{\pi}\,\mathrm{erf}(\xi) + B,$$

where $\xi = \eta^{1/2}/2$, $\mathrm{erf}(\xi)$ is the error function and B is a constant.

Now we need to use the boundary condition that $u(0,t) = 0$ for $t > 0$, which translates to $F(0) = 0$. This gives $B = 0$. Finally, we need to use the initial condition that $u(y,0) = U$ for $y > 0$, which translates to $F(\infty) = 1$, and therefore $C = 1/(2\sqrt{\pi})$. Substituting these constants, we have the answer for the velocity field:

$$u(y,t) = U\,\mathrm{erf}(\xi),$$

where $\xi = \frac{1}{2}\left(\frac{y^2}{\nu t}\right)^{1/2}$.

6.3.7   Model solution to question 6.2.7

1. In steady flow

$$D_t\psi = \partial_t\psi + (\mathbf{u} \cdot \nabla)\psi = \partial_y\psi\partial_x\psi - \partial_x\psi\partial_y\psi = 0,$$

so ψ is constant along streamlines.

2. For the absolute value of velocity we have:

$$u = \sqrt{u_x^2 + u_y^2} = \sqrt{(\partial_y\psi)^2 + (\partial_x\psi)^2} = |\nabla\psi|.$$

3. By splitting the curve into n small sub-intervals and finding the mass flux S_i on each sub-interval i as $\mathbf{u}_i \times d\mathbf{l}_i$ we have (see figure 6.9):

$$S = \lim_{n\to\infty} \sum_{i=1}^{n} S_i = \int_{\mathbf{x}_1}^{\mathbf{x}_2} \mathbf{u} \times d\mathbf{l} = \int_{\mathbf{x}_1}^{\mathbf{x}_2} (\nabla\psi \times \mathbf{e}_z) \times d\mathbf{l} =$$

$$\int_{\mathbf{x}_1}^{\mathbf{x}_2} \nabla\psi \cdot d\mathbf{l} = \psi(\mathbf{x}_2) - \psi(\mathbf{x}_1).$$

4. The normal velocity component is zero at the boundary, $(\mathbf{u} \cdot \mathbf{n}) = 0$. Therefore $(\nabla\psi)_{\parallel} = 0$.

5. For this flow $\psi = \frac{1}{2}\ln(x^2 + y^2)$. The streamlines $\psi = $ const are concentric circles: they are shown in figure 6.10. This is a point vortex flow. It is irrotational everywhere except at the origin $\mathbf{x} = 0$.

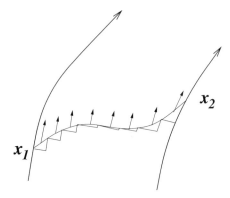

FIGURE 6.9: Calculation of the mass flux.

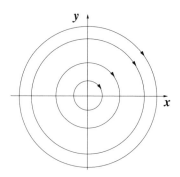

FIGURE 6.10: Point vortex.

6.3.8 Model solution to question 6.2.8

1. From the incompressibility condition, we find that the radial component of the velocity field is zero everywhere in physical space. Thus, the velocity and the vorticity gradient are perpendicular to each other, $\mathbf{u} \cdot \nabla \omega = 0$. Therefore the stationary vorticity equation is satisfied.

2. For the velocity circulation over a circular contour with radius r we have:

$$2\pi r u(r) = \pi r^2 \kappa \quad \text{for } r < a, \quad \text{and} \quad 2\pi r u(r) = \pi a^2 \kappa \quad \text{for } r > a.$$

Thus, for the velocity field we have the following expressions,

$$u(r) = \frac{r\kappa}{2} \quad \text{for } r < a, \quad \text{and} \quad u(r) = \frac{a^2\kappa}{2r} \quad \text{for } r > a.$$

Now let us find the pressure starting with the vortex exterior, $r > a$. In this part the flow is irrotational and we can use Bernoulli's theorem:

$$p = p_0 - \frac{\rho u^2}{2} = p_0 - \frac{a^4 \kappa^2 \rho}{8r^2},$$

where p_0 is the pressure at $r \to \infty$.

Let us now consider the vortex interior, $r < a$. Since the motion in this part is identical to the solid body rotation, we will used a method similar to the one we use in question 3.3.3. The steady-state Euler equation in component form is:

$$(\mathbf{u} \cdot \nabla)u = -\partial_x p / \rho,$$
$$(\mathbf{u} \cdot \nabla)v = -\partial_y p / \rho,$$

or

$$x\kappa^2/4 = -\partial_x p / \rho,$$
$$y\kappa^2/4 = -\partial_y p / \rho.$$

Integrating, we have

$$p = \frac{\kappa^2 \rho}{8}(x^2 + y^2) + C,$$

where C is a constant which is fixed by the matching at $r = a$: $C = p_0 - \frac{a^2 \kappa^2 \rho}{4}$.

3. Taking the limit $a \to 0$ while keeping the combination $\Gamma = \kappa a^2$ constant, we have:

$$u(r) = \frac{\Gamma}{2\pi r}$$

and

$$p(r) = p_0 - \frac{\Gamma^2 \rho}{8r^2}.$$

6.3.9 Model solution to question 6.2.9

Consider a two-dimensional potential flow described by complex potential

$$w(z) = Cz^n,$$

1. For $n = 1$ we have $w(z) = Cz$, i.e. $u - iv = \partial_z w = C$. Therefore, this case corresponds to a flow with a uniform velocity $\mathbf{u} = (C, 0)$.

2. For $n = 2$ we have $w(z) = Cz^2$, i.e. $u - iv = \partial_z w = 2Cz = 2C(x + iy)$. Therefore, this case corresponds to a uniform-strain flow with strain value $\alpha = 2C$.

3. We have

$$w = \phi + i\psi = Cz^n = Cre^{in\theta},$$

i.e.

$$\phi = Cr\cos(n\theta), \quad \text{and} \quad \psi = Cr\sin(n\theta).$$

For the smallest and the second-smallest positive angles for which $\psi = 0$, we have $\theta_1 = 0$ and $\theta_2 = \pi/n$.

4. Recall that lines $\psi = $ const are streamlines. Therefore, the radial rays $\theta = \theta_1$ and $\theta = \theta_2$ (on which $\psi = 0$) are streamlines, i.e. there is no velocity component normal to these rays/planes. In the other words, the free-slip boundary conditions are satisfied at these planes.

5. Sketches of the flows arising in the cases $n = 4$, $n = 4/3$, $n = 2/3$ and $n = 1/2$ are given in figures 6.11, 6.12, 6.13 and 6.14 respectively.

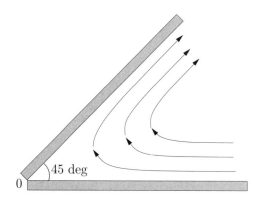

FIGURE 6.11: Case $n = 4$.

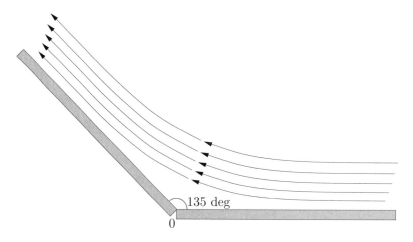

FIGURE 6.12: Case $n = 4/3$.

6. For the absolute value of the velocity field we have $|\mathbf{u}| = |\partial_z w| = nCr^{n-1}$. Thus, for $r \to 0$ the velocity tends to zero for cases with $n < 1$ and it tends to infinity for $n > 1$.

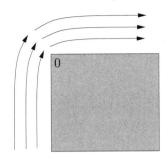

FIGURE 6.13: Case $n = 2/3$.

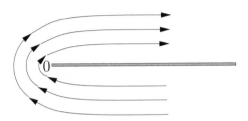

FIGURE 6.14: Case $n = 1/2$.

6.3.10 Model solution to question 6.2.10

1. When C is real positive (negative)—this is a source (sink) flow whose velocity field has only radial component: $v_r = C/r$. When C is purely imaginary, then this is a point vortex with circulation $\Gamma = -2\pi C$. It has only azimuthal velocity $v_\theta = -C/r$.

2. Consider a 2D potential flow described by the complex potential

$$w(z) = \ln z - (1 + ia) \ln(z - 1),$$

where a is a real positive constant.

This is a flow with a source at $z = 0$ and a sink combined with a point vortex at $z = 1$. The stagnation point $z = z_0$ is found from the condition of zero velocity, i.e.

$$\partial_z w(z) = 1/z - (1 + ia)/(z - 1) = 0,$$

which gives $z_0 = i/a$.

3. Because the source at $z = 0$ has the same strength as the sink at $z = 1$, their contributions will asymptotically cancel each other as $r = |z| \to \infty$. So the remaining flow will be the one of the point vortex of circulation $\Gamma = 2\pi a$ with purely azimuthal component $v_\theta = a/r$.

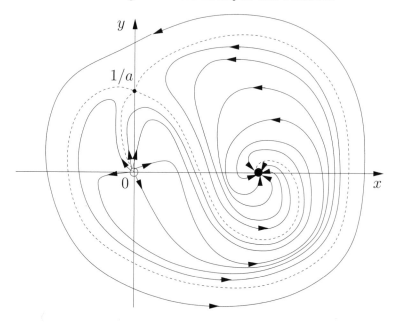

FIGURE 6.15: "Binary star system".

4. A sketch of the streamlines is shown in figure 6.15. The separatrices are shown by the dashed lines. The stream lines originating at the source at $z = 0$ terminate at the sink at $z = 1$. Because of the vortex component, the stream lines near $z = 1$ look like spirals. At large distances the streamlines are closed and become more and more circular as $|z| \to \infty$.

5. In the case $a = 0$ the source-sink system is symmetric, and the quickest way from the source to the sink is along the straight line along the x-axis. The time between the emergence of a fluid particle at the source and its disappearance into the sink can be found by solving the equation

$$\dot{x} = u = \frac{1}{x} - \frac{1}{x-1},$$

which gives

$$-x(x-1)\,dx = dt,$$

or, integrating, we have

$$t = \left[-\frac{1}{3}x^3 + \frac{1}{2}x^2 \right]_0^1 = \frac{1}{6}.$$

6.3.11 Model solution to question 6.2.11

1. Taking the gradient of the potential (6.22) we have:

$$u_x = \partial_x \phi = -Cke^{ky} \sin(kx - \omega t), \quad u_y = \partial_y \phi = Cke^{ky} \cos(kx - \omega t).$$

2. The stream function ψ is defined via

$$u_x = \partial_y \psi = -Cke^{ky} \sin(kx - \omega t), \quad u_y = -\partial_x \psi = Cke^{ky} \cos(kx - \omega t).$$

Integrating we have:

$$\psi = -Ce^{ky} \sin(kx - \omega t).$$

The complex potential is then:

$$\xi = \phi + i\psi = Ce^{ky} \left[\cos(kx - \omega t) - i\sin(kx - \omega t)\right] =$$
$$Ce^{ky - i(kx - \omega t)} = Ce^{-i(kz - \omega t)},$$

where

$$z = x + iy.$$

3. A particle which in the still fluid was at position (x, y), in presence of the wave will move along a trajectory around (x, y):

$$(x(t), y(t)) = (x, y) + (\tilde{x}(t), \tilde{y}(t)).$$

For the particle trajectories we have

$$\dot{\tilde{x}}(t) = u_x = -Cke^{ky} \sin(kx - \omega t), \tag{6.33}$$
$$\dot{\tilde{y}}(t) = u_y = Cke^{ky} \cos(kx - \omega t). \tag{6.34}$$

Integrating these equations we have:

$$\tilde{x}(t) = -\frac{Ck}{\omega} e^{ky} \cos(kx - \omega t), \tag{6.35}$$

$$\tilde{y}(t) = -\frac{Ck}{\omega} e^{ky} \sin(kx - \omega t). \tag{6.36}$$

Thus these trajectories are circles:

$$x^2 + y^2 = R^2,$$

where

$$R = \frac{Ck}{\omega} e^{ky}.$$

The streamlines are given by the equation

$$\psi = -Ce^{ky} \sin(kx - \omega t) = \text{const}.$$

They are infinite periodic curves with parts going to $y \to \pm\infty$. The particle trajectories are different from the streamlines, which is natural for a non-stationary motion.

6.3.12   Model solution to question 6.2.12

1. Near the wing surface, velocity is nearly tangential to the surface and its variation is in the normal direction. Therefore the vorticity is parallel to the wing. The vorticity field is solenoidal i.e. vortex lines cannot end within the fluid. Therefore they must continue beyond the wing tips, so that the vorticity flux remains the same.

   Therefore, the circulation of the trailing vortices is the same as the circulation around the wing.

2. To keep the plane flying, the gravity force must be balanced by the lift, so
$$Mg = \rho V \Gamma (2L)$$
   or
$$\Gamma = \frac{Mg}{2\rho V L}.$$

3. If the vortices are thin and straight, then each of them produces velocity like a point vortex in 2D, namely $\frac{\Gamma}{2\pi r}$ where r is the distance from the vortex. In the middle point, the velocities produced by the two vortices add up: $U = \frac{\Gamma}{2\pi (D/2)} \cdot 2$. Thus,
$$U = \frac{\Gamma}{\pi L}.$$

6.3.13   Model solution to question 6.2.13

1. The physical unit for force, e.g. the drag and lift forces, is the Newton, which is equal to $kg * m/s^2$. Therefore, force per unit length has the physical dimension of $[\rho]L^3/T^2$.

2. The only dimensional combination of U, ρ and L having the dimension of the force per unit length is
$$F = c_1 \rho U^2 L,$$
   where c_1 is a dimensionless constant.

   The only dimensional combination of U, ρ and L having the dimension of the velocity circulation is
$$\Gamma = c_2 U L,$$
   where c_2 is a dimensionless constant.

   From the above two expressions we find Zhukovskiy's lift force:
$$F = c\rho U \Gamma,$$
   where c is a dimensionless constant. More rigorous analysis (Zhukovskiy's theorem) yields $c = 1$.

3. For the drag force per unit length D (similarly to the lift force) assuming that the relevant dimensional quantities determining this quantity are U, ρ and L, we find

$$D = c\rho U^2 L,$$

where c is a dimensionless constant (which, in this case is not necessarily equal to one: it is called the drag coefficient and it strongly dependent on the body shape).

4. The most general expression for the drag force assuming that the relevant dimensional quantities determining this quantity are U, ρ, ν and L (using the result of the previous question and using Re instead of ν) is

$$D = c\rho U^2 L \, f(Re),$$

where c is dimensionless constant and f is a dimensionless function.

Since, according to D'Alembert's paradox, the drag must vanish in the limit $\nu \to 0$, we conclude that $f(Re)$ must be a decreasing function of Re for laminar flows. For turbulent flows the situation is more difficult, because of the flow separation such flows do not tend to the ideal flow solutions in the limit $\nu \to 0$.

6.3.14 Model solution to question 6.2.14

1. Integrating equation (6.23) over y, we have for the individual terms:

$$\int_{-\infty}^{+\infty} u u_x \, dy = \frac{1}{2} \partial_x \int_{-\infty}^{+\infty} u^2 \, dy,$$

$$\int_{-\infty}^{+\infty} v u_y \, dy = -\int_{-\infty}^{+\infty} v_y u \, dy = \int_{-\infty}^{+\infty} u_x u \, dy = \frac{1}{2} \partial_x \int_{-\infty}^{+\infty} u^2 \, dy,$$

and

$$\nu \int_{-\infty}^{+\infty} \partial_{yy} u \, dy = u_y \big|_{-\infty}^{+\infty} = 0.$$

Therefore we find:

$$M = \int_{-\infty}^{+\infty} u^2 \, dy = C,$$

where C is an independent of x constant.

2. Substituting the self-similar solution into the expression for M we have:

$$M = \int_{-\infty}^{+\infty} u^2 \, dy = x^{2a+b} \int_{-\infty}^{+\infty} (f')^2 \, d\eta \,.$$

But M must be independent of x, so

$$b = -2a.$$

3. This leads to the following expressions for the velocity components and their derivatives:

$$u = \psi_y = x^{-a} f',$$
$$v = -\psi_x = ax^{a-1}(-f + 2\eta f'),$$
$$u_x = -ax^{-a-1}(f' + 2\eta f''),$$
$$u_y = x^{-3a} f'',$$
$$u_{yy} = x^{-5a} f'''.$$

Substituting these expressions into equation (6.23) we have:

$$-x^{-a} f' ax^{-a-1}(f' + 2\eta f'') + ax^{a-1}(-f + 2\eta f')x^{-3a} f'' = \nu x^{-5a} f'''.$$

This equation must be independent of x, so

$$a = 1/3 \quad \text{and} \quad b = -2/3.$$

4. The required equation immediately follows from the above equation upon substituting $a = 1/3$.

   The boundary conditions are:

   - $y = 0$ is a streamline, therefore $f(0) = 0$.
   - $u(-y) = u(y) \implies u_y = 0$ at $y = 0$, therefore $f''(0) = 0$.
   - $u|_{y=\infty} = 0$, therefore $f'(\infty) = 0$.

6.3.15 Model solution to question 6.2.15

1. For the vorticity we have $\boldsymbol{\omega} = (0, 0, \omega)$, where $\omega = -\nabla^2 \psi$.

2. The velocity, the stream function and the vorticity of the flow uniformly rotating with angular velocity $\boldsymbol{\Omega}$ are:

   $$\mathbf{u} = \boldsymbol{\Omega} \times \mathbf{x}, \ \psi = -\tfrac{\Omega}{2}(x^2 + y^2), \ \boldsymbol{\omega} = 2\,\boldsymbol{\Omega}.$$

3. We have the following equation for the vorticity:

   $$\partial_t \omega + \mathbf{u} \cdot \nabla \omega = 0.$$

   The left-hand side here is the time derivative of vorticity along the fluid paths, $D_t \, \omega$, therefore this equation describes conservation of vorticity along trajectory of each fluid particle.

4. The vorticity field will remain constant, as in the uniformly rotating flow, $\boldsymbol{\omega} = 2\,\boldsymbol{\Omega} = (0, 0, 2\Omega)$.

5. The fact that the vorticity is constant, together with equation $\omega = -\nabla^2\psi$, suggests that ψ will be a quadratic function of x and y: like in part 2, but now with $\psi = $ const at the elliptic boundary. Therefore, at $t > 0$ we have:

$$\psi = -\frac{\Omega a^2 b^2}{a^2 + b^2}[(x/a)^2 + (y/b)^2],$$

where we found the pre-factor so that $-\nabla^2\psi = 2\Omega$.

6.3.16   Model solution to question 6.2.16

1. For a thin layer, $\partial_x h \ll 1$, the flow is mostly in the x-direction. Thus, from the y-component of the momentum equation we have $\partial_y p = 0$, i.e. p is independent of y. But at the water surface the pressure is atmospheric, $p = p_0$, so $p = p_0$ everywhere in the fluid (it is independent of x and y).

2. Consider a small volume of fluid between x and $x + dx$ and with a unit length in the transverse to x direction: $dx \times h(x) \times 1$. Let $F(x)$ be the momentum flux,

$$F = \rho h u^2.$$

The gravity force onto the volume element is:

$$\mathbf{F}_g = m\mathbf{g} = \rho h \, dx \, \mathbf{g}.$$

The momentum input per unit time by the rain is:

$$M = \dot{m}V_x = (Q \, dx \, \cos\theta)(V \sin\theta).$$

(We have the x-momentum only because the y-momentum is completely dissipated by the roof).

Thus, the momentum balance for the considered small volume is:

$$F(x + dx) - F(x) \approx dx \, \partial_x F = F_{gx} + M,$$

or

$$\rho \partial_x(hu^2) = \rho g h \sin\theta + QV \sin\theta \cos\theta.$$

For the mass balance we have:

$$dx\partial_x(\rho hu) = \text{rain input} = Qdx \cos\theta,$$

or

$$\rho \partial_x(hu) = Q \cos\theta.$$

Integrating this equation we have:

$$\rho hu = Qx \cos\theta.$$

3. The gravity force can be ignored if

$$\rho g h \sin\theta \ll QV \sin\theta \cos\theta,$$

or

$$\rho g h \ll QV. \tag{6.37}$$

In this case, integrating the momentum balance equation we have:

$$\rho h u^2 = QV x \sin\theta \cos\theta + C,$$

where C is a constant. Combining this result with the mass equation, we get:

$$u = V \sin\theta \quad \text{and} \quad h = \frac{Q\cos\theta}{\rho V \sin\theta}x,$$

where we put $C = 0$ since u must be finite at $x = 0$. Using this solution one can rewrite the condition (6.37) as

$$V^2 \gg gx.$$

6.3.17   Model solution to question 6.2.17

1. It is clear from figure 6.6 that the flow is plane parallel:

$$\mathbf{u}(x, y) = (u(y), 0), \tag{6.38}$$

where function $u(y)$ is to be found.

The no-slip boundary condition in this case is

$$u(0) = 0, \tag{6.39}$$

and the pressure boundary condition at the free surface is

$$p(x, h) = p_0. \tag{6.40}$$

2. Stress at the free surface is given by the tangential to the surface component of the viscous stress tensor,

$$\sigma_{ij} = \rho\nu(\partial_{x_i} u_j + \partial_{x_j} u_i),$$

namely by $\sigma_{xy} = \rho\nu\partial_y u(y)$. Absence of the stress at the free surface therefore means that

$$\partial_y u = 0 \quad \text{at} \quad y = h.$$

3. It is clear that the pressure field in this problem is independent of x. Thus, the stationary Navier-Stokes equations written in components are:

$$\frac{1}{\rho}\partial_y p + g \cos\theta \quad = \quad 0, \tag{6.41}$$

$$\nu\partial_{yy}u + g \sin\theta \quad = \quad 0. \tag{6.42}$$

Integrating equation (6.41) and using the pressure boundary condition (6.40), we have

$$p = p_0 + \rho g(h - y)\cos\theta. \tag{6.43}$$

Integrating equation (6.42) twice, and using the no-slip boundary condition (6.39) and the condition $\partial_y u = 0$ at $y = h$, we have

$$u = \frac{g\sin\theta}{2\nu}y(2h - y). \tag{6.44}$$

4. The mass of fluid passing through the cross-section of the layer, per unit length in the transverse direction, per unit time is

$$\int_0^h \rho u\, dy = \frac{\rho g h^3 \sin\theta}{3\nu}. \tag{6.45}$$

Chapter 7

Point vortices and point sources

7.1 Background theory

Point vortex is a 2D flow generated by a singular vorticity distribution which is concentrated at a single point, such that the velocity circulation along a contour embracing this point has a finite value Γ. We have already met this solution in questions 6.2.7 and 6.2.8. In question 6.2.8 we obtained the point vortex solution as a limiting case of a Rankine vortex with vorticity $\Omega(\mathbf{x})$ which is constant inside a circle of radius a, $\Omega(\mathbf{x}) = \kappa = \text{const}$, and zero outside. Then, the limit $a \to 0$ was taken while keeping the circulation constant, $\Gamma = \pi a^2 \kappa = \text{const}$. This means that the vorticity value inside the circle tends to infinity, $\kappa \sim 1/a^2$. In the limit, such a vorticity field will be represented by a Dirac delta function,

$$\Omega(\mathbf{x}) = \Gamma \, \delta(\mathbf{x}). \tag{7.1}$$

This vortex generates the following velocity field,

$$u = -\frac{\Gamma}{2\pi} \frac{y}{x^2 + y^2}, \quad v = \frac{\Gamma}{2\pi} \frac{x}{x^2 + y^2}; \tag{7.2}$$

see problem 6.2.7. Indeed, for this flow the circulation is zero for all contours which do not encircle the vortex, and equal to Γ for all contours encircling the vortex (please show that!). In question 6.2.7 you were also asked to find the stream function for the point vortex flow.

Now we will consider a more general situation when there are $N \geq 2$ point vortices in an ideal 2D flow, each having its own circulation; see figure 7.1. Each vortex in the set will produce a velocity field at the positions of all the other vortices, which will make them move. Remarkably, such motion will change neither the fact that the vortices are point-like, nor the number of vortices, nor the circulation of each individual vortex. This follows from vorticity conservation along fluid paths in 2D ideal flows which we discussed in chapter 2 section 2.1.7. One has to think of applying these results to the situation where vorticity is bounded within vortex patches of tiny areas (points in the limit) separated by distances which are much greater than the size of each of these vortex patches. Indeed, assuming that the vortex patches remain bounded in size (which is true for sufficiently strong vortices separated by

sufficiently large distances), the vorticity of each vortex patch will be moved
with the velocity field made out of the contributions of all the other vortex
patches, as well as by the self-induced velocity the patch produces within
itself. The later contribution can be ignored at long times because the vortex
self-induced velocity just rotates the fluid particles around the centre of the
vortex and does not contribute to the displacement of the vortex patch to
large distance from its original position. Obviously, because the vorticity of
each fluid particle is conserved along its path, the velocity circulation around
each of the vortex patches will also be conserved.

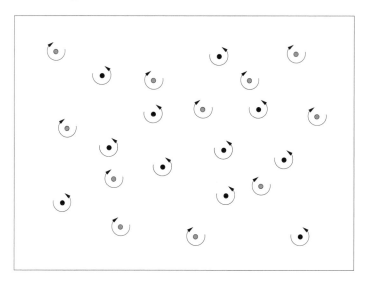

FIGURE 7.1: "Gas" of point vortices. Counter-clockwise and clockwise
pointing arrows mark vortices with positive and negative circulations respectively.

Thus, in the limit of small radii of the vortex patches one can consider
them as a set (or "gas") of N point vortices each having a Dirac-delta vorticity
distribution at an evolving position $\mathbf{x}_j(t)$ ($j = 1, ..., N$) and having its own
circulation Γ_j:

$$\Omega(\mathbf{x}, t) = \sum_{j=1}^{N} \Gamma_j \, \delta(\mathbf{x} - \mathbf{x}_j(t)). \tag{7.3}$$

So the problem of finding the flow is now reduced to finding the trajectories
$\mathbf{x}_j(t)$ ($j = 1, ..., N$) from the condition that the j-th vortex velocity $\dot{\mathbf{x}}_j(t)$
is equal to the sum of the velocity contributions produced by all the other
vortices at the position $\mathbf{x}_j$. Velocity produced by the k-th vortex at the point
$\mathbf{x}_j$ can be found using formula (7.2) in which we shift $\mathbf{x} \to \mathbf{x}_j - \mathbf{x}_k$. The sum
of all $N - 1$ vortex contributions gives us an equation for the rate of change

of $\mathbf{x}_j$:

$$\dot{\mathbf{x}}_j = \sum_{k=1,k\neq j}^{N} \frac{\Gamma_k}{2\pi} \frac{\hat{\mathbf{z}} \times (\mathbf{x}_j - \mathbf{x}_k)}{|\mathbf{x}_j - \mathbf{x}_k|^2}, \tag{7.4}$$

where $\hat{\mathbf{z}}$ is the unit vector normal to the plane of motion, and sign $\times$ stands for the cross product (prove this formula!).

Introducing a complex coordinate for the vortex labelled j as $z_j(t) = x_j(t) + iy_j(t)$, we can rewrite equation (7.4) in a complex form:

$$\frac{dz_j^*}{dt} = -\frac{i}{2\pi} \sum_{k=1,k\neq k}^{N} \frac{\Gamma_k}{z_j - z_k}, \tag{7.5}$$

where star $*$ means complex conjugation.

Another kind of point object arising in the study of ideal fluid flows are point sources or sinks. They have only a radial (with respect to the position of the source/sink) component of velocity, $v(r)$, which is independent of the angular coordinates:

$$\mathbf{u} = \hat{\mathbf{r}}v(r), \tag{7.6}$$

where $\hat{\mathbf{r}}$ is a unit vector pointing away from the source/sink in the radial direction; see figure 7.2.

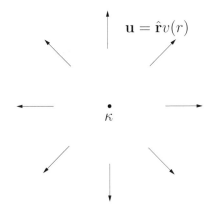

FIGURE 7.2: A point source flow.

From incompressibility we have $r^{d-1}v(r) = \text{const}$, where d is the dimension of the system (which is 2 or 3) and r is the distance from the source/sink. Therefore

$$v(r) = \frac{\kappa}{r^{d-1}}, \tag{7.7}$$

where constant κ is called the source strength when it is positive and the sink strength when it is negative. The constant κ is proportional to the mass flux out of the source or into the sink respectively.

The source/sink flow is irrotational for $r > 0$ and its velocity potential is $\phi(r) = \int v\, dr$ i.e.

$$\phi(r) = \kappa \ln r, \quad \text{or} \quad \phi(r) = -\frac{\kappa}{r}. \tag{7.8}$$

7.2   Further reading

Further discussion of point vortex systems can be found in the books *Hydrodynamics* by H. Lamb [13], *An Introduction to Fluid Dynamics* by G.K. Batchelor [4], *Elementary Fluid Dynamics* by D.J. Acheson [1]. An advanced discussion of various point vortex equilibrium configuration can be found in the review by H. Aref et al. [2]. A discussion of using the point-vortex gas models for describing 2D turbulence can be found in a review by R. H. Kraichnan and D. Montgomery *Two-Dimensional Turbulence* [11] as well as in the original 1949 paper by L. Onsager, who pioneered this approach [18].

7.3   Problems

7.3.1   Energy, momentum, and angular momentum of a point vortex set

Consider a point vortex set described by equations (7.3).

1. Prove that the dynamics of such a vortex gas conserves the total point-vortex interaction energy defined by the formula

$$E_{\mathrm{PV}} = -\frac{1}{4\pi} \sum_{j,k=1;k\neq j}^{N} \Gamma_j \Gamma_k \ln |\mathbf{x}_j - \mathbf{x}_k|. \tag{7.9}$$

2. Prove that the dynamics also conserves the total momentum given by the formula

$$\mathbf{P} = (P_x, P_y) = \sum_{j=1}^{N} \Gamma_j \mathbf{x}_j. \tag{7.10}$$

3. Prove that the dynamics also conserves the total angular momentum given by the formula

$$M = \sum_{j=1}^{N} \Gamma_j |\mathbf{x}_j|^2. \tag{7.11}$$

7.3.2 Motion of two point vortices

> Given information: conservation laws for the energy and momentum (7.9) and (7.10).

Consider a system of two point vortices with circulations Γ_1 and Γ_2 separated by a distance d from each other at $t = 0$.

1. Find the distance between the vortices at $t > 0$.

2. Find the trajectory of each vortex for $\Gamma_1 \neq -\Gamma_2$. How do the vortices move if $\Gamma_1 > 0$ and $\Gamma_2 > 0$? If $\Gamma_1 > 0$ and $\Gamma_2 < 0$? If $\Gamma_1 = \Gamma_2 > 0$?

3. Find the motion when $\Gamma_1 = -\Gamma_2$.

7.3.3 Vortex "molecules"

In this problem we will consider "vortex molecules" composed of several identical point vortices arranged in a symmetric regular polygon configuration. Here we are not concerned about stability of our molecules (it was shown by J.J. Thomson that they are stable for the number of vortices equal to seven or less).

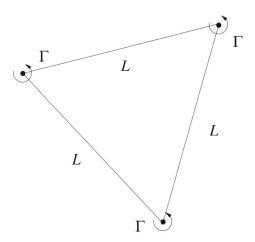

FIGURE 7.3: Three-vortex of molecule.

1. Consider three point vortices with equal circulations $\Gamma > 0$ placed initially (at $t = 0$) at the vertices of an equilateral triangle with sides of length L; see figure 7.3. Find the motion of this vortex system for $t > 0$.

2. Now consider four point vortices with equal circulations $\Gamma > 0$ placed initially at the vertices of a square with sides of length L; see figure 7.4. Find the motion of this vortex system.

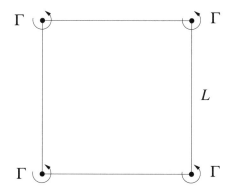

FIGURE 7.4: Four-vortex of molecule.

3. Let us now generalise this setup to an arbitrary number N of vortices with equal circulations $\Gamma > 0$ placed initially at the vertices of a regular polygon (a polygon that has all sides equal and all interior angles equal) with sides L; see figure 7.5 for the $N = 6$ example. Describe qualitatively how such an N-vortex molecule moves (you do not need to find the precise formulas in this part).

4. In our N-vortex molecule considered in the previous part, let $\Gamma = a/N$ where a is a fixed positive real constant. Find the motion of such a molecule in the limit $N \to \infty$ assuming that the vortices lie on a circle with fixed radius R.

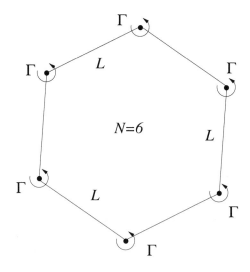

FIGURE 7.5: N-vortex of molecule for $N = 6$.

7.3.4 Motion of three point vortices

> Given information: expressions (7.9), (7.10) and (7.11) for the energy, the momentum and the angular momentum invariants of the point vortex system considered in question 7.3.1.

The system of three point vortices appears to be fully integrable for any vortex circulations and initial vortex positions. This is because (as shown by Poincaré) the system has three integrals of motion in involution which are needed for integrability by Liouville's theorem: the energy, the angular momentum and the square of the modulus of the momentum. However, the motion of three point vortices in general may be rather complicated, and here we will restrict ourselves with a simpler special case.

Let us consider a system of three point vortices, two of which have positive circulations Γ and the third one with a negative circulation $-\Gamma$. The negative vortex is initially separated from one of the positive vortices by a distance d which is much less than the distance to another positive vortex; see figure 7.6. Thus initially the positive-negative vortex pair does not feel the presence of the third vortex and propagates along almost a straight line as a vortex dipole (c.f. problem 7.3.2). However, the direction of the dipole propagation is toward the third vortex and, upon approaching the latter, the dipole's track (shown by the dotted line in figure 7.6) significantly deviates from the straight line: it bends around the third vortex and then moves away from it asymptotically approaching a straight line which is at angle α with its initial direction; see figure 7.6.

1. Explain why the dipole does not feel the presence of the third vortex and propagates along almost in a straight line when the distance d between the vortices making the dipole is much less that the distance from the dipole to a third vortex.

2. Find the distance between the positive and the negative vortex in the dipole after the "collision" with the isolated vortex, i.e. at large time when the dipole has travelled far away from the third vortex.

3. Describe the motion of the third vortex long before and long after its collision with the dipole.

4. Find the total distance **a** travelled by the third vortex (it is indicated by the dashed line in figure 7.6).

7.3.5 Point vortex in a channel

A two-dimensional incompressible inviscid fluid is occupying a channel bounded by two straight walls, $x = 0$ and $x = h$. Consider a flow in such a channel which is generated by a point vortex of circulation $\Gamma > 0$ initially located at the point $(x, y) = (a, 0)$; see figure 7.7.

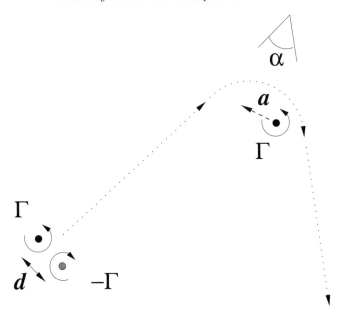

FIGURE 7.6: Scattering of a point vortex dipole at an isolated vortex.

1. Formulate the free-slip boundary conditions for this problem.

2. Explain why the flow generated by such a point vortex in the channel with the free-slip boundary conditions is the same (for $0 < x < h$) as the flow generated by an infinite chain of the image vortices on an unbounded two-dimensional plane which are initially located as shown in figure 7.8.

3. Explain why the point vortex will move parallel to the walls with a constant speed.

4. Find the velocity of the point vortex if it is located in the centre of the channel (i.e. $a = h/2$).

5. By summing the contributions from all of the image vortices, find the velocity of the point vortex for the case $a = h/4$. **Hint**: to evaluate the infinite sum involved use the following identity,

$$\sum_{k=1}^{\infty} \frac{1}{16k^2 - 1} = \frac{1}{2} - \frac{\pi}{8}.$$

6. Find the velocity of the point vortex for the case $a = 3h/4$.

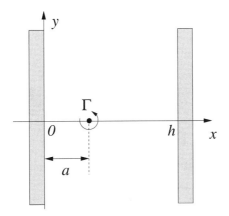

FIGURE 7.7: Vortex in channel.

7.3.6 Point vortices and their images

> Given information: equation for the motion of point vortices expressed in the complex coordinates—equation (7.5).

At some instant $t = 0$, two point vortices, each having circulation $-\Gamma$, are located at $(x_1, y_1) = (a, -a)$ and $(x_2, y_2) = (-a, a)$, and two other point vortices, with circulation Γ, are at $(x_3, y_3) = (a, a)$ and $(x_4, y_4) = (-a, -a)$ on the unbounded plane.

1. Express the complex coordinates of the second, the third and the fourth vortices in terms of the first vortex coordinate z_1 for all times $t > 0$.

2. Find the first vortex's trajectory $y_1(x_1)$ and use the result of part 1 to find trajectories of the other three vortices.

3. Show that the velocity field generated by the four vortices in the upper half-plane $y > 0$ is identical to the one produced by only two vortices, numbers 2 and 3, moving over a fixed plane boundary located at $y = 0$ with a free-slip boundary condition on this plane.

4. When an aircraft takes off, the two vortices that trail from its wingtips are observed to move downwards under each other's influence and then to move farther apart as they approach the ground. Why is this?

7.3.7 Clustering in the gas of point vortices

> Given information: expression (7.9) for the energy of the point vortex system considered in question 7.3.1.

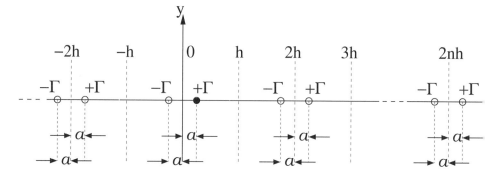

KEY: • = original vortex
       ○ = image vortex

FIGURE 7.8: Infinite chain of image vortices.

In this problem we will deal with a large set of vortices in statistical equilibrium: a system introduced by Lars Onsager in 1940s. We will learn about negative temperature states associated with the grouping of like-signed vortices into large-scale clusters and, therefore, forming a large-scale vortex structure. This process is a close relative of the inverse cascade process in 2D turbulence considered in chapter 8.

Consider a "gas" of point vortices with circulations Γ and $-\Gamma$ in equal numbers; see figure 7.1. Suppose that the vortex system is in a thermodynamic equilibrium which can be qualitatively described by the Boltzmann distribution,

$$P\{\mathbf{x_j}\} \sim e^{-E_{\mathrm{PV}}\{\mathbf{x_j}\}/T},$$

where E_{PV} is the vortex interaction energy which is a function of $\{\mathbf{x_j}\}$—the full set of the vortex coordinates $\mathbf{x_j}$; $j = 1, ..., N$, and $T = \mathrm{const}$ is the temperature. Here, we must postulate that vortices cannot come together closer than some minimal distance d which is much less than the mean intervortex distance: otherwise the total energy would be unbounded, and the Boltzmann distribution would not be normalisable. (The minimal distance d may represent a qualitative re-introduction of the effect of the vortex core diameter which was neglected in the point vortex approximation).

1. Show that for sufficiently large positive temperatures T, the most probable states could be viewed as a gas of tight dipoles, i.e. it consists of pairs of positive and negative vortices whose distance from each other is much less than the distance to the other vortices.

2. Show that for negative temperatures T, the most probable states contain large-scale clusters such that vortices of one sign dominate within each cluster.

7.3.8 Discharge through a hole

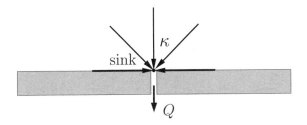

FIGURE 7.9: Water discharging through a hole at the bottom.

A flow of water is discharging through a hole at the flat bottom of a large flume; see figure 7.9.

1. Explain why the resulting water flow is equivalent to the flow produced by a point sink of strength $\kappa < 0$ assuming that the free-slip boundary condition at the wall is satisfied.

2. Find κ in terms of the volume flow Q through the hole.

3. Now suppose that the water is pumped into the flume through the same hole rather than being discharged. Explain why it would be less realistic to consider the resulting flow as a point source flow.

7.3.9 Submerged pump near a wall

A pump is submerged deep under the water surface at an initial distance L_0 from a vertical wall; see figure 7.10. The pump is suspended vertically so that it could freely move with the fluid flow in the horizontal direction. We will assume that the effects of the bottom and of the free surface can be neglected.

1. Explain why the flow produced by the pump with the free-slip boundary condition at the wall is equivalent to the flow produced by a point sink with strength $\kappa < 0$ and its image sink with the same strength $\kappa < 0$ as shown in figure 7.10. Find κ in terms of the volume flow Q through the pump.

2. Assuming that the pump moves freely with the fluid flow in the horizontal direction, find the distance from the wall as a function of time, $L(t)$, and the time t^* at which the pump will hit the wall.

7.3.10 Flows past a zeppelin and a balloon

Flows past blunt bodies of revolution, e.g. a zeppelin-shaped body; see figure 7.11, can be constructed by combining a uniform flow $\mathbf{u} = U\hat{\mathbf{x}}$ with a

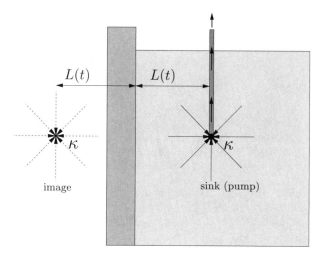

FIGURE 7.10: Submerged pump near a wall.

set of point sources and sinks on the x-axis (here $\hat{\mathbf{x}}$ is a unit vector along the x-axis).

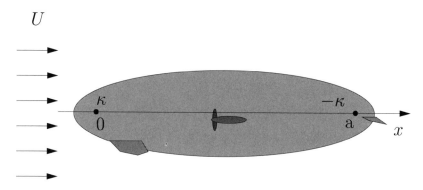

FIGURE 7.11: Flow around a zeppelin.

1. Let us first study a case with only one source and no sinks. Consider a three-dimensional flow whose velocity potential is a sum of the individual contributions corresponding to a uniform flow with velocity U along the x-axis and a source of strength κ at $\mathbf{x} = (0, 0, 0)$.

   Find the velocity field of such a flow.

2. Find the stagnation point at which the velocity is zero.

3. Describe properties of the streamline passing through the stagnation point. How does this streamline behave far downstream of the stagnation

point? Explain why this streamline corresponds to the shape of the body of revolution to which the considered flow is a solution with the free-slip boundary condition.

4. Now let us consider a case with only one source and one sink. Namely, let the velocity potential be a sum of the individual contributions corresponding to a uniform flow with velocity U along the x-axis, a source of strength κ at $\mathbf{x} = (0,0,0)$ and a sink of the same strength κ located at $\mathbf{x} = (a,0,0)$; see figure 7.11.

Find the velocity field of such a flow.

5. Consider the limit $a \to 0$ keeping $\mu = \kappa a$ constant. Find the velocity potential ϕ and the radial component of velocity u_r. Find the surface where $u_r = 0$. What is the shape of the body of revolution in the limit $a \to 0$? (**Hint:** think of the free-slip boundary condition in terms of u_r.)

7.4   Solutions

7.4.1   Model solution to question 7.3.1

1. Let us differentiate the total point-vortex energy defined by formula (7.9):

$$
\dot{E}_{\text{PV}} = -\frac{1}{4\pi} \sum_{j,k=1;\, k\neq j}^{N} \Gamma_j \Gamma_k \frac{(\dot{\mathbf{x}}_j - \dot{\mathbf{x}}_k) \cdot (\mathbf{x}_j - \mathbf{x}_k)}{|\mathbf{x}_j - \mathbf{x}_k|^2} =
$$
$$
-\frac{1}{2\pi} \sum_{j,k=1;\, k\neq j}^{N} \Gamma_j \Gamma_k \frac{\dot{\mathbf{x}}_j \cdot (\mathbf{x}_j - \mathbf{x}_k)}{|\mathbf{x}_j - \mathbf{x}_k|^2}. \tag{7.12}
$$

Substituting $\dot{\mathbf{x}}_j$ from the equation of motion (7.4), we have

$$
\dot{E}_{\text{PV}} = -\frac{1}{4\pi^2} \sum_{j,k,m=1;\, k,m\neq j}^{N} \Gamma_j \Gamma_k \Gamma_m \frac{\hat{\mathbf{z}} \cdot [(\mathbf{x}_j - \mathbf{x}_m) \times (\mathbf{x}_j - \mathbf{x}_k)]}{|\mathbf{x}_j - \mathbf{x}_m|^2 |\mathbf{x}_j - \mathbf{x}_k|^2}. \tag{7.13}
$$

The expression under the sum is antisymmetric with respect to the change $k \leftrightarrow m$, so $\dot{E}_{\text{PV}} = 0$.

2. Differentiating the total momentum given by formula (7.10) and using formula (7.4), we have:

$$
\dot{\mathbf{P}} = \sum_{j=1}^{N} \Gamma_j \dot{\mathbf{x}}_j = \sum_{j,k=1, k\neq j}^{N} \frac{\Gamma_j \Gamma_k}{2\pi} \frac{\hat{\mathbf{z}} \times (\mathbf{x}_j - \mathbf{x}_k)}{|\mathbf{x}_j - \mathbf{x}_k|^2}. \tag{7.14}
$$

The expression under the sum is antisymmetric with respect to the change $j \leftrightarrow k$, so $\dot{\mathbf{P}} = 0$.

3. Differentiating the total angular momentum given by the formula (7.11) and using formula (7.4), we have:

$$\dot{M} = \sum_{j=1}^{N} 2\Gamma_j \, \mathbf{x}_j \cdot \dot{\mathbf{x}}_j = \sum_{j,k=1,k\neq j}^{N} \frac{\Gamma_j \Gamma_k}{\pi} \frac{\mathbf{x}_j \cdot (\hat{\mathbf{z}} \times (\mathbf{x}_j - \mathbf{x}_k))}{|\mathbf{x}_j - \mathbf{x}_k|^2}. \qquad (7.15)$$

But $\mathbf{x}_j \cdot (\hat{\mathbf{z}} \times (\mathbf{x}_j - \mathbf{x}_k)) = -\mathbf{x}_j \cdot (\hat{\mathbf{z}} \times \mathbf{x}_k) = \mathbf{x}_k \cdot (\hat{\mathbf{z}} \times \mathbf{x}_j)$ and, therefore, $\dot{M} = 0$.

7.4.2 Model solution to question 7.3.2

1. Distance between the vortices at $t > 0$ is the same as the distance at $t = 0$. This follows from the energy conservation, which for the system of two vortices is

$$-\frac{\Gamma_1 \Gamma_2}{4\pi} \ln|\mathbf{x}_1 - \mathbf{x}_2| = \text{const.}$$

2. First consider the case $\Gamma_1 \neq -\Gamma_2$.

From the momentum conservation we obtain that the "mass centre" of the two-vortex system is fixed:

$$\mathbf{x}^* = \frac{\Gamma_1 \mathbf{x}_1 + \Gamma_2 \mathbf{x}_2}{\Gamma_1 + \Gamma_2} = \text{const.}$$

But the above two expressions also mean that

$$d_1 = |\mathbf{x}_1 - \mathbf{x}^*| = \frac{|\Gamma_2||\mathbf{x}_1 - \mathbf{x}_2|}{|\Gamma_1 + \Gamma_2|} = \text{const.}$$

and

$$d_2 = |\mathbf{x}_2 - \mathbf{x}^*| = \frac{|\Gamma_1||\mathbf{x}_1 - \mathbf{x}_2|}{|\Gamma_1 + \Gamma_2|} = \text{const.}$$

From this we conclude that both vortices move in circles with the same angular velocity and the same centre at $\mathbf{x} = \mathbf{x}^*$.

The modulus of the angular velocity Ω can be found by dividing the vortex one's velocity modulus, $|v_1| = |\Gamma_2|/(2\pi d)$, by the distance to the rotation centre, d_1. We have

$$|\Omega| = \frac{|\Gamma_1 + \Gamma_2|}{2\pi d^2}.$$

The sign of Ω is the same as the sign of $\Gamma_1 + \Gamma_2$, so finally

$$\Omega = \frac{\Gamma_1 + \Gamma_2}{2\pi d^2}.$$

If $\Gamma_1 > 0$ and $\Gamma_2 > 0$, the vortices move counter-clockwise around the common centre which is in between of the two vortices; see figure 7.12. If $\Gamma_1 = \Gamma_2 > 0$, then the centre of rotation is exactly in the middle of the piece of line connecting the two vortices.

If $\Gamma_1 > 0$ and $\Gamma_2 < 0$, then the rotation centre $\mathbf{x}^*$ is located outside of the piece of line connecting the vortices; see figure 7.13.

3. In the case $\Gamma_1 = -\Gamma_2 = \Gamma$, both vortices produce exactly the same velocity at each other's location, $v_1 = v_2 = \Gamma/(2\pi d)$. Thus, the vortex pair will move together perpendicular to the line connecting the vortices with velocity

$$v = \frac{\Gamma}{2\pi d}.$$

This structure is called a vortex dipole.

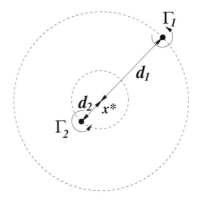

FIGURE 7.12: Motion of two like-signed point vortices. Both vortices are counter-clockwise in this figure, i.e. they have positive circulations $\Gamma_1 > 0$ and $\Gamma_2 > 0$.

7.4.3 Model solution to question 7.3.3

1. Velocity of vortex 3 will be equal to the sum of velocities produced by vortices 1 and 2 at the position of vortex 3, $\mathbf{V}_{13}$ and $\mathbf{V}_{23}$ respectively; see figure 7.15. Note that $\mathbf{V}_{13}$ and $\mathbf{V}_{23}$ have equal absolute values, and the absolute value of their sum will be equal to the sum of projections of V_{13} and V_{23} onto the line bisecting their directions (the dashed line in figure 7.15):

$$V = |\mathbf{V}| = 2|\mathbf{V}_{13}| \cos \theta = \sqrt{3} \frac{\Gamma}{2\pi L} = \frac{\Gamma}{2\pi R},$$

where R is the radius of the circle on which the vortices lie.

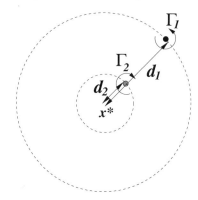

FIGURE 7.13: Motion of two opposite-signed point vortices. Counterclockwise and clockwise pointing arrows mark vortices with positive and negative circulations respectively. In this example $\Gamma_1 > 0$ and $\Gamma_2 < 0$.

Because of the symmetry, the velocity of each of the other two vortices will have the same absolute value V and it will also be directed tangentially to the circle on which the vortices lie. Thus, the molecule will experience a solid body rotation with angular velocity

$$\Omega = V/R = \frac{\Gamma}{2\pi R^2}. \tag{7.16}$$

2. Similarly for the four-vortex molecule, the velocities of all the four vortices will have the same absolute values and they will all be tangential to the circle on which the vortices lie. Thus again, the molecule will experience a solid body rotation with constant angular velocity. The vortex velocity is made up of the three velocity contributions produced at the location of the vortex by the other three vortices:

$$V = 2\frac{\Gamma}{2\pi L}\cos\frac{\pi}{4} + \frac{\Gamma}{2\pi 2R} = \frac{3\Gamma}{4\pi R}.$$

Correspondingly, for the angular velocity we have

$$\Omega = V/R = \frac{3\Gamma}{4\pi R^2}. \tag{7.17}$$

3. From the symmetry, for an N-vortex with arbitrary N the motion will be similar: it will rotate as a solid body with a constant angular velocity. From formulae (7.16) and (7.17) you could guess that in this case

$$\Omega = \frac{(N-1)\Gamma}{4\pi R^2} \tag{7.18}$$

(although you are not asked find this formula in this part).

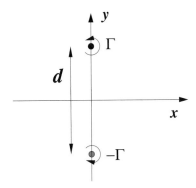

FIGURE 7.14: Dipole of point vortices. Counter-clockwise and clockwise pointing arrows mark vortices with positive and negative circulations respectively.

4. In the N-vortex molecule let $\Gamma = a/N$ and take the limit $N \to \infty$ with radius R fixed. Easy to see that this limit corresponds to a uniform distribution of vorticity on the circle and, therefore, to an axially symmetric flow without radial component of velocity. The azimuthal velocity in such a flow changes from zero at $r < R$ to

$$V_{r>R} = \frac{N\Gamma}{2\pi r} = \frac{a}{2\pi r}$$

for $r > R$.

For the vortices on the circle $r = R$, velocity is equal to the middle of the jump value,

$$V = \frac{V_{r=R+0} - V_{r=R-0}}{2} = \frac{a}{4\pi R},$$

and

$$\Omega = V/R = \frac{a}{4\pi R^2}.$$

7.4.4 Model solution to question 7.3.4

1. The pair of vortices separated by distance d produce, at each other's positions, velocities which are equal to each other and much greater (by factor r/d) than the velocity induced by the third vortex separated from the dipole vortices by distance $r \gg d$. Thus the vortex dipole moves along almost a straight line with velocity $\Gamma/(2\pi d)$.

2. After the collision with the isolated vortex, i.e. at large time when the dipole has travelled far away from the third vortex, the distance between the positive and the negative vortex in the dipole will return to its initial

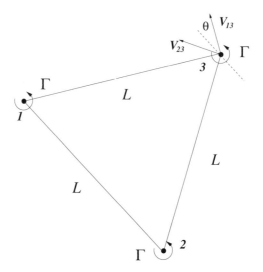

FIGURE 7.15: Three-vortex of molecule.

value d. This follows from the energy conservation (7.9), because the dominant energy contribution long before and long after the collision will be from the dipole pair:

$$E_d = \frac{\Gamma^2}{4\pi} \ln d \approx \text{const.}$$

3. The position of the third vortex long before and long after its collision with the dipole is almost constant because the velocity produced at its position by the dipole's vortices is negligible. In fact, because of the leading-order cancellations of the velocity contributions of the two vortices within the dipole, this velocity scales as $\sim \Gamma d/r^2$ i.e. it decays more rapidly with distance r than it would in the case of a single vortex velocity.

4. When the dipole is close to the third vortex, the latter moves. From the momentum conservation law (7.10), from which (cancelling Γ) we have:

$$(\mathbf{x}_2 - \mathbf{x}_1 + \mathbf{x}_3)|_{t=0} = (\mathbf{x}_2 - \mathbf{x}_1 + \mathbf{x}_3)|_{t=+\infty}$$

The total distance travelled by the third vortex, $\mathbf{a} = \mathbf{x}_3|_{t=+\infty} - \mathbf{x}_3|_{t=0}$, is then

$$\mathbf{a} = \left(2d\cos\frac{\alpha}{2}\right)\hat{\mathbf{n}},$$

where $\hat{\mathbf{n}}$ is the unit vector normal to the line bisecting the initial and the final directions of the dipole motion; see figure 7.6.

7.4.5   Model solution to question 7.3.5

1. Let the velocity field be $\mathbf{u} = (u(x, y, t), v(x, y, t))$. The free-slip boundary conditions for this problem are:

$$u(x, 0, t) = u(x, h, t) = 0; \quad v(x, 0, t) \ \& \ v(x, h, t) \ \text{—arbitrary.}$$

2. From symmetry, the flow generated by an infinite chain of the image vortices on an unbounded two-dimensional plane which are located as shown in figure 7.8 will have $u(x, 0, t) = u(x, h, t) = 0$, as required by the free-slip boundary conditions of the original problem. Thus, the rest of the flow for $0 < x < h$ will also be the same, because it contains the same point vortex inside this interval and it satisfies the fluid equations (basically the Laplace equation for the velocity potential of an irrotational incompressible flow) outside of the point vortex.

3. The point vortex and its images will remain on the same line for any $t > 0$. Thus, they will not produce any x-component of the velocity at each other's positions. Because the x-positions of the vortices do not change, they produce the same y-component of the velocity at each other's positions for any $t > 0$. Therefore, the point vortex (and its images) will move parallel to the walls with a constant speed.

4. From the symmetry, the velocity of the point vortex located in the centre of the channel (when $a = h/2$) is zero.

5. Here we have to refer to figure 7.8. By summing the contributions from all of the image vortices for the case $a = h/4$, we have for the vortex velocity V:

$$V = \frac{\Gamma}{2\pi h} \sum_{k=-\infty}^{k=+\infty} \left[\frac{1}{2k - 1/2} - \frac{1}{2k + 1/2} \right] =$$

$$-\frac{\Gamma}{\pi h} + \frac{4\Gamma}{\pi h} \sum_{k=1}^{k=+\infty} \frac{1}{16k^2 - 1} = \frac{\Gamma}{\pi h}(1 - \pi/2).$$

6. From the symmetry, the velocity of the point vortex for the case $a = 3h/4$ is just the negative of the velocity found for the case $a = h/4$,

$$V = -\frac{\Gamma}{\pi h}(1 - \pi/2).$$

7.4.6   Model solution to question 7.3.6

1. From the symmetry of the problem for $t \geq 0$ we have

$$x_1 = x_2 = -x_3 = -x_4; \quad y_1 = -y_2 = -y_3 = y_4.$$

Thus, we have the following expressions for the complex coordinates of the 2nd, 3rd and the 4th vortices in terms of the 1st vortex coordinate z_1: From the symmetry of the problem for $t \geq 0$ we have

$$z_1 = -z_2 = z_3^* = -z_4^*,$$

where star * stands for complex conjugate.

2. From the equation for the motion (7.5), after substituting the above relations, we have the following equation for the coordinate of the first vortex,

$$\dot{z}_1^* \equiv \dot{x}_1 - i\dot{y}_1 = \frac{i\Gamma}{2\pi} \left[\frac{1}{2z_1} - \frac{1}{z_1 - z_1^*} - \frac{1}{z_1 + z_1^*} \right] =$$

$$\frac{i\Gamma}{2\pi} \left[\frac{x_1 - iy_1}{2(x_1^2 + y_1^2)} - \frac{1}{2iy_1} - \frac{1}{2x_1} \right].$$

Taking the ratio of the real and imaginary parts of this equation we have

$$\frac{\dot{y}_1}{\dot{x}_1} = -\frac{y_1^3}{x_1^3}.$$

Integrating this equation we obtain the trajectories of the 1st vortex

$$\frac{1}{y_1^2} - \frac{1}{x_1^2} = C = \text{const.} \tag{7.19}$$

For each fixed constant C, this equation describes four disconnected branches (in the different quadrants of the (x, y)-plane), each corresponding to a trajectory of one of the four vortices. The 1st vortex will move in the lower-right quadrant.

3. By the symmetry, the vertical velocity of the four-vortex flow is zero at $y = 0$, which means that the free-slip boundary condition on this plane is satisfied for the two-vortex system. For the latter flow the other two vortices at $y < 0$ become the image vortices.

4. The aircraft's trailing vortices move downward as a vortex dipole, and upon approaching closer to the ground they behave like the 2nd and the 3rd vortices in our problem. According to our solution (7.19), this corresponds to decreasing y and increasing x, i.e. the vortices move apart as they approach the ground.

7.4.7   Model solution to question 7.3.7

For the purposes of this problem, we shift expression (7.9) for the energy of the point vortex system by a constant reference energy:

$$E_{\text{PV}} = -\frac{1}{4\pi} \sum_{j,k=1;k\neq j}^{N} \Gamma_j \Gamma_k \ln \frac{|\mathbf{x}_j - \mathbf{x}_k|}{\ell}, \tag{7.20}$$

where ℓ is a typical inter-vortex distance, $\ell = 1/\sqrt{n}$, n being the number vortices per unit area. This shift does not change the form of the Boltzmann distribution,

$$P\{\mathbf{x_j}\} \sim e^{-E_{\mathrm{PV}}\{\mathbf{x_j}\}/T},$$

since the proportionality constant would have to be normalised anyway.

1. According to expression (7.20), the energy of the point vortex system consists of a sum of pairwise contributions: pairs of like-signed and oppositely signed vortices make positive and negative contributions respectively when they are tight, $|\mathbf{x}_j - \mathbf{x}_k| < \ell$, and vice versa if they are distant, $|\mathbf{x}_j - \mathbf{x}_k| > \ell$.

   For the vortex system with Boltzmann distribution the negative energy states are more probable for positive temperatures T. For large temperatures, the most probable states are those that correspond to the vortex configurations with largest possible energies, i.e. such that they consist of tight vortex dipoles with sizes $\ll \ell$, namely opposite-signed pairs with distance between the vortices of the order of the minimal allowed inter-vortex distance d.

2. For negative temperatures T, the most probable states are configurations with like-signed vortex pairs that have on average smaller inter-vortex distances than the opposite-signed pairs. Such distributions can be viewed as large-scale clusters such that vortices of one sign dominate within each cluster. For very large negative temperatures T the most probable state consists of just two such clusters: a positive and a negative, each being tight, with distances between the vortices within each cluster of the order of the minimal distance d.

7.4.8 Model solution to question 7.3.8

1. The velocity field produced by a sink located at the wall will be tangential to this wall, i.e. the free-slip boundary condition at the wall is satisfied.

2. The radial component of the velocity field produced by the sink of strength κ in the 3D case is

$$v(r) = \frac{\kappa}{r^2}, \tag{7.21}$$

and the inward volume flux through a sphere of radius r surrounding the sink is $-4\pi r^2 v(r) = -4\pi\kappa$. However, in the problem only half of the sink flow is realised in the fluid volume and, therefore, the volume flow through the hole is $Q = -2\pi\kappa$. Thus, for κ in terms of the volume flow through the pump Q we have:

$$\kappa = -\frac{Q}{2\pi}. \tag{7.22}$$

3. For the water discharging from the same hole we would most likely encounter the flow separation phenomenon. In the other words, the flow would emerge out of the hole as a jet, which is strongly anisotropic and, therefore, very dissimilar to the point source flow, which is isotropic. In a much more viscous fluid such a flow separation would be suppressed, but the free-slip boundary condition would have to be replaced with the no-slip condition, and the isotropic flow would be again impossible.

7.4.9   Model solution to question 7.3.9

1. The flow produced by a sink with strength $\kappa < 0$ and its image sink with the same strength and at the same distance on the other side (as shown in figure 7.10) will produce zero normal component of velocity at the wall—as required by the free-slip boundary conditions. Note the difference with the point vortices: the image has the same sign and not the opposite one as in the point vortex case.

   The radial component of the velocity field produced by the sink of strength κ in the 3D case is

   $$v(r) = \frac{\kappa}{r^2}, \qquad (7.23)$$

   and the inward volume flux through a sphere of radius r surrounding the sink is

   $$Q = -4\pi r^2 v(r) = -4\pi\kappa. \qquad (7.24)$$

   Thus, for κ in terms of the volume flow through the pump Q we have:

   $$\kappa = -\frac{Q}{4\pi}. \qquad (7.25)$$

2. From equation (7.23) we have that the flow velocity produced by the image sink at the pump's position is:

   $$v_i = \frac{\kappa}{4L^2}. \qquad (7.26)$$

   The self-induced velocity of the sink at the pump is zero because its velocity distribution is spherically symmetric (c.f. with the fact that the self-induced velocity for the point vortex is also zero). Therefore, the equation of motion for the pump drifting with the flow is:

   $$\dot{L} = \frac{\kappa}{4L^2}, \qquad (7.27)$$

   solving which we have:

   $$\frac{1}{3}(L^3 - L_0^3) = \frac{\kappa}{4}(t - t_0), \qquad (7.28)$$

or

$$L(t) = \left[\frac{3\kappa}{4}(t - t_0) + L_0^3 \right]^{1/3}. \tag{7.29}$$

Note that because $\kappa < 0$, the distance to the wall is decreasing: $L(t) < L_0$ for $t > t_0$. The pump hits the wall at the moment $t = t^*$ determined by the condition $L(t^*) = 0$, which gives

$$t^* - t_0 = -\frac{3\kappa L_0^3}{4} = \frac{3QL_0^3}{16\pi}. \tag{7.30}$$

7.4.10   Model solution to question 7.3.10

1. The velocity potential is

$$\phi = Ux - \frac{\kappa}{\sqrt{x^2 + y^2 + z^2}}. \tag{7.31}$$

The velocity field is axially symmetric with respect to the x-axis without an azimuthal component, and, therefore, it is sufficient to consider points on the (x, y)-plane only (i.e. for $z = 0$). In this plane we have

$$\mathbf{u} = \nabla\phi = \left(U + \frac{\kappa x}{(x^2 + y^2)^{3/2}} \right) \hat{\mathbf{x}} + \kappa y \left(\frac{1}{(x^2 + y^2)^{3/2}} \right) \hat{\mathbf{y}}. \tag{7.32}$$

2. At the stagnation points we have $\mathbf{u} = 0$, so from the y-component (z-component) of the velocity we get $y = 0$ ($z = 0$) and for the x-component we have $x = -\sqrt{\kappa/U}$. So, we have the following stagnation point:

$$\mathbf{x}_s = \left(-\sqrt{\frac{\kappa}{U}}, 0, 0 \right). \tag{7.33}$$

3. The streamline passing through the stagnation point is the shape of the body of revolution to which the considered flow is a solution satisfying the free-slip boundary condition, because the velocity is always tangent to the streamline. Other streamlines do not correspond to a blunt body shape, because any flow round a blunt body must have a stagnation point.

At $x \to \infty$ we have $\mathbf{u} \to U\hat{\mathbf{x}}$, and the shape of the body of revolution tends to a cylinder parallel to the x-axis. Radius R of this cylinder can be obtained from equating the mass production rate at the source, $4\pi\kappa$, to the mass flux carried downstream within the circle of radius R, i.e. $U\pi R^2$. This gives

$$R = 2\sqrt{\frac{\kappa}{U}} = 2|\mathbf{x}_s|.$$

4. The velocity potential is

$$\phi = Ux - \frac{\kappa}{\sqrt{x^2 + y^2 + z^2}} + \frac{\kappa}{\sqrt{(x-a)^2 + y^2 + z^2}}. \tag{7.34}$$

The velocity field is axially symmetric with respect to the x-axis and, therefore, it is sufficient to consider points on the (x, y)-plane only (i.e. for $z = 0$). In this plane we have

$$\mathbf{u} = \nabla\phi = \left(U + \frac{\kappa x}{(x^2 + y^2)^{3/2}} - \frac{\kappa(x-a)}{((x-a)^2 + y^2)^{3/2}}\right)\hat{\mathbf{x}}$$

$$\kappa y\left(\frac{1}{(x^2 + y^2)^{3/2}} - \frac{1}{((x-a)^2 + y^2)^{3/2}}\right)\hat{\mathbf{y}}. \tag{7.35}$$

5. In the limit $a \to 0$, using a Taylor expansion in a we have to the leading order:

$$\phi \approx Ux + \frac{\mu x}{r^3}, \tag{7.36}$$

where $r = \sqrt{x^2 + y^2 + z^2}$.

Putting again $z = 0$ and substituting $x = r\cos\theta$, for the radial component of velocity we have

$$u_r = \partial_r\phi = Ur\cos\theta - \frac{2\mu\cos\theta}{r^3} = \frac{Ux}{r} - \frac{2\mu x}{r^4}, \tag{7.37}$$

and we have $u_r = 0$ on a sphere with radius b given by

$$b = \left(\frac{2\mu}{U}\right)^{1/3}.$$

Thus we conclude that the ideal flow with velocity U past a sphere of radius b is given by the potential

$$\phi = Ux\left(1 + \frac{b^3}{2r^3}\right). \tag{7.38}$$

Chapter 8

Turbulence

8.1 Background theory

Turbulence is a state in which fluid particles move chaotically. Even though the fluid equations are deterministic, turbulent motion of individual fluid particles is unpredictable due to an intrinsic instability of such a motion, i.e. great sensitivity of the fluid paths to slight disturbances. In this, turbulence is similar to a gas of molecules: even if one uses deterministic laws of classical mechanics to describe the motion and collisions of the molecules, their individual tracks are impossible to predict. However, in both the turbulence theory and the kinetic theory of gases one deals with *averaged* quantities which are possible to describe and predict, such as the mean concentration of molecules in gas, or a mean velocity profile distribution (in both gas and turbulence), or the mean velocity distribution of gas molecules, or a *mean distribution of energy over the turbulent eddies.*

The latter is quantified by the *energy spectrum* which is the central object in the turbulence theory. The one-dimensional kinetic energy spectrum of three-dimensional turbulence is defined via the Fourier transform of a velocity correlator as

$$\hat{E} = \frac{k^2}{(2\pi)^2} \int \langle \mathbf{u}(\mathbf{x}) \cdot \mathbf{u}(\mathbf{x} + \mathbf{r}) \rangle e^{i\mathbf{k}\cdot\mathbf{r}} \, d\mathbf{r}, \tag{8.1}$$

where $\mathbf{k}$ is the wave vector and $k = |\mathbf{k}|$ and $\langle \rangle$ means the ensemble average. The spectrum $\hat{E}(k)$ describes distribution of the energy over the length scales $l = 2\pi/k$ (roughly, over the eddies with diameters l).

Important idealisations in the turbulence theory are assumptions of turbulence *homogeneity* and *isotropy*, respectively, when all positions and all directions in the physical space are statistically equivalent. Obviously, to obey these properties turbulent motion must occupy an infinite space, which may only be an approximation to real turbulent flows. In homogeneous turbulence spectrum $\hat{E}$ is independent of the position $\mathbf{x}$; in isotropic turbulence $\hat{E}$ is independent of the direction of $\mathbf{k}$, i.e. $\hat{E} \equiv \hat{E}(k)$.

Previously we mentioned an analogy between turbulence and a gas of particles referring to the unpredictability of individual particle trajectories in both cases. However, there is a crucial difference between these two systems: gases are often found in states close to thermodynamic equilibrium described by some temperature T, whereas turbulence is a strongly non-equilibrium state. In turbulence, large eddies are generated by an external forcing (e.g. stirring) or by an instability. Such large eddies do not live long: their mutual interactions lead to breaking of these vortices into smaller ones. If viscosity is small enough (i.e. Reynolds number is big) all of the energy of the initial large eddies will be transferred to the smaller eddies without loss. But the smaller eddies will also interact and will further break to even smaller eddies passing to them the turbulent energy. The process of breakup will continue until energy reaches very small scales at which the viscous dissipation is important, and at these scales the kinetic energy of the turbulent motion will be transferred to the internal energy. This occurs at the scales with sizes such that Reynolds number estimated at these length scales is of order one.

The picture described above was introduced in the 1920s by Richardson and it is called the energy cascade through scales. It is illustrated in figure 8.1. When energy is supplied to the system at a constant rate, a statistically steady state is expected to form with a time independent energy spectrum $\hat{E}$.

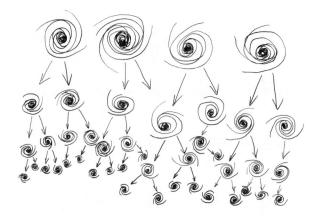

FIGURE 8.1: Richardson cascade in the physical space.

Using the rule of correspondence between the length scale and the wave vector, $l = 2\pi/k$, one can represent the turbulent cascade of energy as a step-by-step energy transfer from smaller to larger wave vector magnitudes as schematically shown in figure 8.2. The range of k's in which there are no sources or sinks of the turbulent energy and where the energy is transferred between adjacent scales without change is called the *inertial range*. It is clear that in the steady state the flux of energy ϵ, i.e. the rate at which it is transferred from smaller to larger k's, is independent of k and is equal to the rate at which it is produced for the forcing at the largest scales, which in turn is equal to the rate of viscous dissipation at the smallest scales of the turbulent motion.

The central result in the theory of turbulence is the Kolmogorov energy

Energy cascade

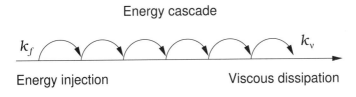

Energy injection   Viscous dissipation

FIGURE 8.2: Richardson's cascade in the k-space.

spectrum, which can be obtained for the inertial range of scales from a simple dimensional argument which expresses the spectrum $\hat{E}$ in terms of the dissipation rate of the kinetic energy ϵ and the wave vector magnitude k. This argument will be considered in question 8.3.1.

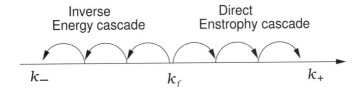

FIGURE 8.3: 2D turbulence: dual cascade in the k-space space.

Let us now consider two-dimensional (2D) turbulence; the 1D energy spectrum in this case is:

$$\hat{E} = \frac{k}{4\pi} \int \langle \mathbf{u}(\mathbf{x}) \cdot \mathbf{u}(\mathbf{x} + \mathbf{r}) \rangle e^{i\mathbf{k} \cdot \mathbf{r}} \, d\mathbf{r}. \tag{8.2}$$

In 2D turbulence, in addition to the energy there exists another positive and quadratic (with respect to velocity) invariant, the enstrophy; see equation (2.25). Presence of two conserved positive quadratic quantities results in a dual cascade behaviour, namely the energy will cascade toward smaller wave vectors (larger scales) and the enstrophy will cascade toward larger wave vectors (smaller scales). Because the energy cascade transfer in this case is opposite to the one in 3D turbulence, such an energy cascade is called *inverse energy cascade*. Respectively, the enstrophy cascade is referred to as a *direct cascade*. Such conclusions about the cascade directions are usually established by an *ad absurdum* Fjørtoft argument. This argument for steady (forced and dissipated) 2D turbulence will be considered in question 8.3.2. For evolving 2D turbulence, in systems without forcing and dissipation, the energy and the enstrophy transfer directions can be found using the Cauchy-Schwartz inequality; this will be considered in question 8.3.3.

Energy spectra corresponding to the inverse energy cascade and the forward enstrophy cascade can be obtained dimensionally, in a way similar to the Kolmogorov spectrum derivation in 3D turbulence. They are called Kraichnan spectra and they will be considered in question 8.3.4.

Interestingly, a Kolmogorov-type scaling was obtained in turbulence theory long before Kolmogorov found it in his spectrum. Richardson discovered it in 1920s by considering experimental data and finding a law of separation of two tracer particles in 3D turbulence, which will be considered in question 8.3.5.

Finally, we will consider turbulence which arises in a fluid flow close to a solid boundary in question 8.3.6. Such near-wall turbulence is inhomogeneous and anisotropic, and we will, following von Kármán, aim to find the mean velocity profile using dimensional considerations.

8.2 Further reading

A detailed account of the turbulence theory is given in the books *Turbulence: The Legacy of A.N. Kolmogorov* by U. Frisch [8] and *The Theory of Homogeneous Turbulence* by G.K. Batchelor [3]. Some facts and approaches in turbulence are described in the books *Fluid Dynamics* by L.D. Landau and E.M. Lifshitz [14] and *Elementary Fluid Mechanics* by T. Kambe [9]. An excellent source of information on the 2D turbulence is a review by R. H. Kraichnan and D. Montgomery *Two-Dimensional Turbulence* [11]. The reader is also recommended to see the original 1941 paper by A.N. Kolmogorov [10], which is very clearly written and accessible to non-specialists.

8.3 Problems

8.3.1 Kolmogorov spectrum of turbulence

1. Find the physical dimensions of the energy spectrum $\hat{E}(k)$ and of the dissipation rate of the kinetic energy ϵ.

2. Use a dimensional argument to find the kinetic energy spectrum assuming that the only dimensional quantities it can depend on are ϵ and k.

8.3.2 Dual cascade in steady two-dimensional turbulence

Given information: the energy and the enstrophy densities in the two-dimensional plane, E and Z respectively, are expressed in terms of the one-dimensional energy spectrum $\hat{E}(k)$ as follows:

$$E = \int_0^\infty \hat{E}(k)\, dk, \quad \text{and} \quad Z = \int_0^\infty k^2\, \hat{E}(k)\, dk,$$

where $k = |\mathbf{k}|$ is the wavevector length.

In this problem we will consider a statistically steady state of forced and dissipated 2D turbulence and, following Fjørtoft an *ad absurdum* argument [7] resulting in the dual cascade picture illustrated in figure 8.3.

Consider 2D turbulence which is forced at wave numbers k_f and dissipated in two regions, $k < k_-$ and $k > k_+$ such that $k_- \ll k_f \ll k_+$.

1. Find the relation between the energy production rate ϵ and the enstrophy production rate η.

2. Suppose by contradiction that in steady-state conditions energy is dissipated at wave numbers $k > k_+$ at a rate comparable to the energy production rate ϵ. Find a lower bound for the enstrophy dissipation rate in this case, and show that this bound is greater than the enstrophy production rate η. Thus conclude that this is impossible in steady-state conditions and, therefore, most of the energy has to be dissipated at $k < k_-$.

3. Find a similar contradiction argument to show that most of the enstrophy has to be dissipated at $k > k_+$.

8.3.3   Dual cascade in evolving two-dimensional turbulence

Given information: the energy and the enstrophy densities in the two-dimensional plane, E and Z respectively, are expressed in term of the one-dimensional energy spectrum $\hat{E}(k)$ as follows:

$$E = \int_0^\infty \hat{E}(k)\,dk, \quad \text{and} \quad Z = \int_0^\infty k^2\,\hat{E}(k)\,dk,$$

where $k = |\mathbf{k}|$ is the wavevector length.

The energy and the enstrophy centroids describe the wavevectors containing most of energy and enstrophy (at a particular moment of time). They are defined respectively as

$$
\begin{aligned}
k_E &= \int_0^\infty k\,\hat{E}(k)\,dk/E, \\
k_Z &= \int_0^\infty k^3\,\hat{E}(k)\,dk/Z.
\end{aligned}
$$

The Cauchy-Schwarz inequality is

$$\left| \int_0^\infty f(k)g(k)\,dk \right| \le \left| \int_0^\infty f^2(k)\,dk \right|^{1/2} \left| \int_0^\infty g^2(k)\,dk \right|^{1/2}$$

for any functions $f(k), g(k) \in \mathbb{L}^2$.

Usually, Fjørtoft's argument finds the directions of the energy and the enstrophy cascades in *stationary* 2D turbulence in presence of forcing and dissipation. Below, we will consider *evolving* 2D turbulence (no forcing or dissipation), and will re-formulate Fjørtoft's argument in terms of the energy and enstrophy centroids.

1. Explain why the enstrophy is conserved in 2D ideal flows but not in 3D flows.

2. Assuming that the integrals defining E, Z, k_E, k_Z converge and using the Cauchy-Schwarz inequality, prove that:

$$
\begin{aligned}
k_E &\le \sqrt{Z/E}, & (8.3) \\
k_E k_Z &\ge Z/E, & (8.4) \\
k_Z &\ge \sqrt{Z/E}. & (8.5)
\end{aligned}
$$

Hint: You might like to choose $f(k)$ and $g(k)$ from the following list: $\hat{E}^{1/2}, k^{1/2}\hat{E}^{1/2}, k\hat{E}^{1/2}, k^{3/2}\hat{E}^{1/2}$.

3. Interpret inequalities (8.3) and (8.5) in terms of the allowed directions for the energy and the enstrophy transfers in the k-space (cascades). What does inequality (8.4) say about these transfers?

8.3.4 Spectra of two-dimensional turbulence

Given information: the one-dimensional kinetic energy spectrum of two-dimensional turbulence is

$$\hat{E}(k) = \frac{k}{4\pi} \int \langle \mathbf{u}(\mathbf{x}) \cdot \mathbf{u}(\mathbf{x} + \mathbf{r}) \rangle e^{i\mathbf{k}\cdot\mathbf{r}} \, d\mathbf{r},$$

where $\mathbf{k}$ is the wavevector and $k = |\mathbf{k}|$ and $\langle\rangle$ means the ensemble average. Note that the spectrum is independent of the direction of $\mathbf{k}$ because the turbulence is assumed to be isotropic; it is also independent of $\mathbf{x}$ for homogeneous turbulence. The spectrum $\hat{E}(k)$ describes distribution of the energy over the length scales $l = 2\pi/k$.

Consider stationary homogeneous two-dimensional turbulence of an incompressible fluid on an infinite plane which is generated at the length scale l_g. The Kraichnan-Batchelor theory states that, under certain conditions, $\hat{E}(k)$ at $k < 2\pi/l_g$ and at $k > 2\pi/l_g$ is determined only by ϵ and η respectively.

1. Find the physical dimensions of the energy spectrum $\hat{E}(k)$ and of the dissipation rates of the kinetic energy and the enstrophy, ϵ and η.

2. Use dimensional arguments to find the kinetic energy spectrum assuming that the only dimensional quantities it can depend on are ϵ and k.

3. Use dimensional arguments to find the kinetic energy spectrum assuming that the only dimensional quantities it can depend on are η and k.

8.3.5 Dispersion of particles in turbulence

This problem considers the evolution in time of the mean distance δ between two tracer particles (e.g. dust specks) embedded in turbulence. You may know that if such particles experienced Brownian motion then $\delta = D\,t^{1/2}$, where D is a diffusion constant. In turbulence, the mean separation also grows as a power law, $\delta \propto t^n$, but the power n is different from $1/2$. Namely, it is given by the so-called Richardson law which you will be asked to derive in this problem. To do this, you will need to use a dimensional analysis approach similar to the one we have used in question 8.3.1 to derive the Kolmogorov spectrum.

1. Consider 3D turbulence forced at a large scale at which the role of viscous dissipation is negligible. Define the inertial range of scales. Describe in words the Richardson cascade picture.

2. Formulate Kolmogorov's universality hypothesis.

3. Find the physical dimension of the energy injection rate ϵ.

4. Consider two tracer particles which are moving with the fluid particles in 3D turbulence. Assume that the distance between these particles corresponds to the inertial range of scales in turbulence, and use Kolmogorov's universality hypothesis and the dimensional argument to derive Richardson's law

$$\delta = C\,\epsilon^m\,t^n,$$

where C is a dimensionless constant. Find m and n.

5. In his 1926 paper [22], Richardson obtained an equation for $P(\delta, t)$, the probability density function defined so that $P(\delta, t)\,d\delta$ is the probability of the two-particle separation to be in the range from δ to $\delta + d\delta$. He wrote

$$\frac{\partial P}{\partial t} = \frac{1}{\delta^2}\frac{\partial}{\partial \delta}\left(\delta^2 D(\delta)\frac{\partial P}{\partial \delta}\right),$$

where $D(\delta) = C'\epsilon^q\delta^s$ (C' is a dimensionless constant) and he deduced the value of the index s from experimental data, because it was not until 1941 that Kolmogorov's dimensional argument was discovered. For the same reason, the dependence on ϵ was not found (nor even mentioned) by Richardson. Please use the dimensional argument and find q and s.

8.3.6   Near-wall turbulence

In this problem we will, following von Kármán [30], apply a dimensional argument to find mean velocity profile in a turbulent flow near a solid wall.

Consider a flow near an infinite flat boundary located at $y = 0$. Its velocity profile is

$$u(x, y, z, t) = U(y) + \tilde{u}(x, y, z, t),$$

where $U(y)$ is a mean velocity profile and $\tilde{u}$ is velocity field of turbulent pulsations.

Von Kármán's 1930 theory postulates that, far enough from the wall, the momentum flux toward the wall, σ, (the wall friction density) is determined not by the viscous stress (as it would be in a laminar flow) but by the turbulent pulsations which randomly move fluid particles across the mean-flow streamlines.

1. Produce an argument for why it is only the derivative $\partial_y U(y)$ and not the local value of the velocity $U(y)$ that can determine the local properties of the turbulent flow.

2. Find the physical dimensions of $\partial_y U(y)$ and σ.

3. From considering the physical dimensions, find the only possible relation between $\partial_y U(y)$, y and σ assuming that the kinematic viscosity coefficient ν is irrelevant and cannot enter this relation.

4. Integrate the relation you have found and obtain an expression for $U(y)$. This expression will contain a dimensional integration constant which is yet to be found.

5. Now consider the flow which is so close to the wall that the viscosity effect is dominant and the flow is laminar, i.e. $\tilde{u} = 0$. Find $U(y)$ in this region.

6. Assume that the mean value of the momentum flux toward the wall, σ, and the typical turbulent velocity are related via $\sigma \sim \langle \tilde{u}^2 \rangle$. Transition from the turbulent to the viscous layer occurs at a crossover value of y where the mean velocity of the turbulent layer $U(y)$ coincides, up to an order one constant, with the typical turbulent velocity $\sqrt{\sigma}$. Use this condition to eliminate the dimensional constant from your expression for the mean velocity $U(y)$ of the turbulent layer. Your final answer will contain two dimensionless constants which may not be found from our dimensional argument and which are fixed by fitting data for $U(y)$ obtained in experiments.

8.3.7 Dissipative anomaly in turbulence

One of the remarkable properties in turbulence is that the energy dissipation rate tends to a finite limit which is independent of the kinematic viscosity coefficient when the latter tends to zero. Physically, this occurs via a systems response to adjust the scale at which the turbulent energy cascade is dissipated (i.e. the Kolmogorov scale), so that the system can absorb precisely the same amount of energy per unit time as the amount produced at the largest scales by a mechanical forcing or instability.

1. Consider an experiment producing a turbulent von Kármán flow—a setup nicknamed the "French washing machine". In this setup fluid occupies a cylindrical volume and is forced with two counter-rotating propellers—one at the top and the other at the base of the cylinder. The rotation angular velocities are $\pm\Omega$ and the characteristic size of the apparatus is R.

2. Using the dimensional analysis, find a physical estimate for the energy injection rate. How are the energy injection and the dissipation rates related in stationary turbulence?

3. Using the dimensional analysis, find the dissipative scale ℓ_ν in the resulting turbulent state from the condition that the dissipation rate is independent of ν.

8.4 Solutions

8.4.1 Model solution to question 8.3.1

1. The physical dimensions of the energy spectrum $\hat{E}(k)$ can be found from the formula relating it to the kinetic energy density in the physical space:

$$\frac{1}{2}\langle u^2 \rangle = \int \hat{E}(k)\, dk,$$

   where k is the absolute value of the wave vector. From this formula we find that the physical dimension of the energy spectrum $\hat{E}(k)$ is L^3/T^2.

   The dissipation rate of the kinetic energy ϵ is, by definition, the energy dissipated per unit time per unit volume. Thus, the physical dimension of ϵ is L^2/T^3.

2. The only dimensional combination of ϵ and k which has the dimension of the kinetic energy $\hat{E}(k)$ is

$$\hat{E}(k) = C_K \epsilon^{2/3} k^{-5/3},$$

   which is the famous Kolmogorov spectrum [10]. Here C_K is a dimensionless constant which is called the Kolmogorov constant. Experiments indicate that $C_K \approx 0.5$.

8.4.2 Model solution to question 8.3.2

1. The relation between the energy production rate ϵ and the enstrophy production rate η follows from the relation between the k-space densities of the energy and the enstrophy (see the given information). Thus,

$$\epsilon = k_f^2\, \eta.$$

2. Suppose by contradiction that in steady state energy is dissipated at wave numbers $k > k_+$ at a rate ϵ_+ comparable to the energy production rate ϵ. Then we have a lower bound for the enstrophy dissipation in the same region:

$$\eta_+ \geq \epsilon_+ k_+^2 \gg \epsilon_+ k_f^2 \sim \epsilon k_f^2 = \eta.$$

   But it is impossible to dissipate in a steady-state condition more enstrophy per unit time than the amount produced by the forcing. Thus conclude that most of the energy has to be dissipated at $k < k_-$.

3. Similar contradiction argument to show that most of the enstrophy has to be dissipated at $k > k_+$ is left to the reader.

8.4.3 Model solution to question 8.3.3

1. This is because in 2D the vortex stretching term is zero (but not in 3D!), hence the vorticity is conserved along the fluid paths, hence the integrated square vorticity (enstrophy) is conserved.

2. In the Cauchy-Schwartz inequality, we will only deal with positive functions, so the absolute value brackets may be omitted. Being in $\mathbb{L}^2$ in our case means that all the relevant integrals converge, as suggested in the statement of the problem.

   First, let us consider integral $\int k\hat{E}\,dk$ and apply the Cauchy-Schwartz inequality as follows,

   $$\int_0^\infty k\hat{E}\,dk = \int (k\hat{E}^{1/2})(\hat{E}^{1/2})\,dk \leq \left(\int_0^\infty k^2\hat{E}\,dk\right)^{1/2}\left(\int_0^\infty \hat{E}\,dk\right)^{1/2},$$

   which immediately yields (8.3).

   Second, let us consider $\Omega = \int k^2\hat{E}\,dk$ and split it as,

   $$\int_0^\infty k^2\hat{E}\,dk = \int_0^\infty (k^{3/2}\hat{E}^{1/2})(k^{1/2}\hat{E}^{1/2})\,dk \leq$$

   $$\left(\int_0^\infty k^3\hat{E}\,dk\right)^{1/2}\left(\int_0^\infty k\hat{E}\,dk\right)^{1/2},$$

   which immediately yields (8.4).

   Combining (8.4) with (8.3) gives (8.5).

3. Inequalities (8.3) and (8.5) mean that the energy is not allowed to move to high k's and the enstrophy is not allowed to move to low k's. Thus, the energy may move only to low k's—this is the inverse energy cascade, and the enstrophy may move only to high k's—this is the direct enstrophy cascade. However, these inequalities do not oblige the energy and the enstrophy to move in scale—they may remain at the scales where they were put initially (as it is the case e.g. for time-independent 2D solutions).

   On the other hand, condition (8.4) means that if the energy happened to move to low k's, the enstrophy *must* move to high k's.

8.4.4 Model solution to question 8.3.4

1. The physical dimensions of the energy spectrum $\hat{E}(k)$ can be found from its defining formula: $[\hat{E}(k)] = L^{-1}[u]^2L^2 = L^3/T^2$ (this dimension is the same as in the 3D case).

   The dissipation rate of the kinetic energy, ϵ, is, by definition, the energy dissipated per unit time per unit volume. Thus, its physical dimension is $[\epsilon] = [u]^2/T = L^2/T^3$.

The dissipation rate of the enstrophy, η, is, by definition, the enstrophy dissipated per unit time per unit volume. Thus, its physical dimension is $[\eta] = [\omega]^2/T = 1/T^3$.

2. The only dimensional combination of ϵ and k which has the dimension of the kinetic energy $\hat{E}(k)$ is

$$\hat{E}(k) = C_1 \epsilon^{2/3} k^{-5/3},$$

where C_1 is a dimensionless constant. Although this expression is very close to the Kolmogorov energy spectrum, the constant C_1 does not have to coincide with the Kolmogorov constant C_K (indeed, from experiment it appears to be different, $C_1 \approx 7$ and $C_K \approx 0.5$).

3. The only dimensional combination of η and k which has the dimension of the kinetic energy $\hat{E}(k)$ is

$$\hat{E}(k) = C_2 \epsilon^{2/3} k^{-3},$$

where C_2 is a dimensionless constant. This is the Kraichnan spectrum. From numerical experiments $C_2 \approx 2$.

8.4.5 Model solution to question 8.3.5

1. For 3D turbulence forced at a large scale at which the role of viscous dissipation is negligible, energy dissipation by viscosity occurs at a much smaller scale. Starting at the forcing scale, the energy cascades step-by-step from larger to smaller (but comparable in size) scales until it reaches the viscous dissipation scale: this is the Richardson cascade picture. The range of scales in between of the energy injection and the energy dissipation scales, and where the energy balance is determined purely by the energy flux through scales (and not its external injection or dissipation) is called the inertial range.

2. Kolmogorov's universality hypothesis states that all statistical properties of turbulence at any scale within the inertial range are determined by the energy flux only, and independent of the details of the energy injection or dissipation mechanisms (turbulence locality). In stationary turbulence, the energy flux is equal to the injection rate of the kinetic energy density, ϵ.

3. By definition, the energy injection rate ϵ is the kinetic energy density injected into the system in unit time. Thus,

$$[\epsilon] = [u^2]/T = L^2/T^3.$$

4. Let

$$\delta = C\,\epsilon^m\,t^n,$$

where C is a dimensionless constant. The only values of m and n which make this formula dimensionally correct are $m = 1/2$ and $n = 3/2$. This is the famous Richardson 3/2-law of turbulent dispersion.

5. From the PDF equation, $[D] = L^2/T$, so the dimensional analysis gives $D = C'\epsilon^{1/3}\delta^{4/3}$, where C' is a dimensionless constant. So, $q = 1/3$ and $s = 4/3$.

8.4.6 Model solution to question 8.3.6

1. The laws of physics, including the ones of fluid dynamics, are the same in all inertial frames, i.e. the motion will be the same in any two flows whose initial conditions are different by a constant velocity field (up to transformation into the moving frame). Thus, the properties of the turbulent flow, such as the turbulent transport (including the momentum flux across the streamlines) cannot depend on the mean velocity $U(y)$ but only on its derivative $\partial_y U(y)$.

2. The physical dimension of $\partial_y U(y)$ is $1/T$. By the definition of the momentum flux as the amount of the momentum passing through a unit area per unit time, we find the physical dimension of σ to be $[\rho][V][u]/(L^2 T) = L^2/T^2$. Here we choose the system where $\rho = 1$ taking into account the fluid's incompressibility.

   Note that the dimension of σ is the same as the one of squared velocity, which is consistent with the assumption to be made below that $\sigma \sim \langle \tilde{u}^2 \rangle$, where $\tilde{u}$ the typical turbulent velocity.

3. The only possible dimensionally correct relation between $\partial_y U(y)$, y and σ (independent from the kinematic viscosity coefficient ν) is:

$$\partial_y U(y) = \frac{\sqrt{\sigma}}{\kappa y},$$

where κ is a dimensionless number called the Von Kármán constant (it is traditional to define it as a constant in the denominator rather than in the numerator). Integrating this relation, we obtain the following expression for $U(y)$:

$$U(y) = \frac{\sqrt{\sigma}}{\kappa}\ln y + A,$$

where A is a dimensional integration constant which is yet to be found.

4. Since there is no stream-wise pressure gradient, $U(y)$ will be a solution of the x-momentum equation

$$\nu\partial_{yy}U = 0,$$

i.e.

$$U(y) = By,$$

where $B =$ const (we have taken into account the no-slip boundary condition $U(0) = 0$).

There must be the same amount of the x-momentum flux toward the wall through the laminar sub-layer as through the turbulent one, so

$$\nu \partial_y U = \nu B = \sigma.$$

5. Assuming that the transition from the turbulent to the viscous layer occurs at a crossover value $y = y^\dagger$ where the mean velocity of the turbulent layer $U(y)$ matches, up to an order one constant, the typical turbulent velocity $\sqrt{\sigma}$, we have

$$U(y^\dagger) = \frac{\sigma}{\nu} y^\dagger = \sqrt{\sigma} \quad \text{i.e.} \quad y^\dagger = \frac{\nu}{\sqrt{\sigma}},$$

(this is just a definition of $y^\dagger$) and

$$U(y) = \sqrt{\sigma} \left(\frac{1}{\kappa} \ln \frac{y}{y^\dagger} + C \right),$$

where C is an order one constant which remains undetermined by our approximate matching procedure. Together with κ, C is fixed by fitting data for $U(y)$ obtained in experiments, which results in $\kappa \approx 0.4$ and $C \approx 5$.

8.4.7 Model solution to question 8.3.7

1. In stationary turbulence, the energy injection and the dissipation rates are equal on average. By definition, the energy injection (dissipation) rate ϵ is the energy produced (dissipated) in a unit volume per unit time. Thus, the physical dimension of ϵ is $[u^2]/T = L^2/T^3$. The only combination of Ω and R that agrees with this dimension is $R^2\Omega^3$, so

$$\epsilon \sim R^2 \Omega^3.$$

2. The physical dimension of the kinematic viscosity coefficient is L^2/T. To absorb the energy flux ϵ by viscosity, we find (using the dimensional analysis) that

$$\ell_\nu = \frac{\nu^{3/4}}{\epsilon^{1/4}}.$$

Chapter 9

Compressible flow

9.1 Background theory

The compressibility of the fluid becomes essential when the flow velocity is of the same order or greater than the speed of sound (see question 1.3.2) or when the fluid (or gas) is subject to changing external pressure on the containing volume like in an air pump or in a balloon rising through a stratified atmosphere.

In this chapter we will deal with compressible flows described by the inviscid equations of a polytropic gas (1.5), (1.6), (1.7) and (1.8). We will consider both strong and weak solutions of these equations. The latter contain jumps in the fluid density, pressure and velocity fields (shocks, contact discontinuities) which are smoothed when viscosity is taken into account. However, the jump conditions follow from some general conservation laws and they are insensitive to the detailed structure of the transition region.

9.1.1 One-dimensional gas dynamics

We will start by considering classical results for 1D flows for which equations (1.5), (1.6) and (1.7) become:

$$\partial_t u + u \partial_x u = -\frac{1}{\rho} \partial_x p, \tag{9.1}$$

$$\partial_t \rho + \partial_x (\rho u) = 0, \tag{9.2}$$

$$\partial_t S + u \partial_x S = 0. \tag{9.3}$$

A powerful tool for solving the 1D gas equations is the method of characteristics. To use this method one needs to rewrite these equations in the characteristic form:

$$
\begin{aligned}
D_t^+ p + \rho c_s D_t^+ u &= 0, & (9.4)\\
D_t^- p - \rho c_s D_t^- u &= 0, & (9.5)\\
D_t^0 S &= 0, & (9.6)
\end{aligned}
$$

where $c_s = \sqrt{(\partial_\rho p)_S}$ is the speed of sound, $((\partial_\rho \cdot)_S$ means the ρ-derivative with S kept constant), and D_t^+, D_t^- and D_t^0 are the time derivatives along the characteristic curves C^+, C^- and C^0 respectively defined by:

$$
\begin{align}
C^+ &: \quad \dot{x}(t) = u(x(t), t) + c_s(x(t), t), \tag{9.7}\\
C^- &: \quad \dot{x}(t) = u(x(t), t) - c_s(x(t), t), \tag{9.8}\\
C^0 &: \quad \dot{x}(t) = u(x(t), t). \tag{9.9}
\end{align}
$$

Obviously, $D_t^0 \equiv D_t$ and the entropy conservation equation (9.3) is already in the characteristic form. For derivation of the remaining two characteristic equations see problem 9.3.1.

For isentropic flow, $S = $ const, we have $D_t^\pm p = (\partial_\rho p)_S D_t^\pm \rho = c_s^2 D_t^\pm \rho$. In this case one can rewrite equations (9.4) and (9.5) as conservation along the characteristics laws:

$$
\begin{align}
D_t^+ R^+ &= 0, \tag{9.10}\\
D_t^- R^- &= 0, \tag{9.11}
\end{align}
$$

where quantities

$$
R^+ = \int \frac{c_s(\rho)}{\rho} d\rho + u, \tag{9.12}
$$

$$
R^- = \int \frac{c_s(\rho)}{\rho} d\rho - u \tag{9.13}
$$

are called Riemann invariants. For polytropic gases, $p = $ const ρ^γ, hence we have

$$
R^\pm = \frac{2c_s}{\gamma - 1} \pm u, \tag{9.14}
$$

Crossing characteristics of the same family indicates the moment at which the solution with single-valued fields ceases to exist. This happens when disturbances travelling at higher speeds catch up with slower disturbances, and this process is called wave breaking. An example of such breaking solutions is considered in question 9.3.2.

After the wave breaking, viscosity becomes important due to high gradients of the field at the wave breaking location. Sharp transition regions arise and start moving—they are called shock waves or simply shocks. Even though the viscosity is essential in the sharp transition region, the jump conditions relating the flow fields on the two sides of the shock can be determined based on the inviscid conservation laws written in the continuity equation form.

One such continuity equation is (9.2): it describes the mass conservation. Combining equations (9.1) and (9.2), one can also write a continuity equation describing the momentum conservation:

$$
\partial_t(\rho u) + \partial_x(\rho u^2 + p) = 0. \tag{9.15}
$$

Note that the entropy equation (9.3), when combined with the equation (9.2), can also be written as a continuity equation,

$$
\partial_t(\rho S) + \partial_x(\rho u S) = 0. \tag{9.16}
$$

However, the entropy conservation breaks down within shocks where the viscosity is essential, whereas the mass and the momentum conservation persist (see question 9.3.5). Thus, when deriving the jump conditions the entropy continuity equation (9.17) has to be replaced by another conservation law which persists for the weak solutions (the ones with jumps): this appears to be the energy conservation (see question 9.3.6):

$$\partial_t \left(\varepsilon + \frac{\rho}{2}u^2\right) + \partial_x \left[\rho u \left(\frac{u^2}{2} + h\right)\right] = 0, \tag{9.17}$$

where ε is the internal energy density and

$$h = \frac{\varepsilon + p}{\rho} \tag{9.18}$$

is the specific enthalpy. For polytropic gases

$$\varepsilon = \frac{p}{\gamma - 1} \quad \text{and} \quad h = \frac{\gamma p}{(\gamma - 1)\rho} = \frac{c_s^2}{(\gamma - 1)}. \tag{9.19}$$

Relations between the fluid quantities ahead and behind the shock are independent of viscosity in the small viscosity limit. They are found from the flux conservation for the mass, momentum and the total (kinetic plus internal) energy; see equations (9.2), (9.15) and (9.17). In the reference frame where the shock is stationary (not moving), we have:

$$\rho_1 u_1 = \rho_2 u_2, \tag{9.20}$$
$$\rho_1 u_1^2 + p_1 = \rho_2 u_2^2 + p_2, \tag{9.21}$$
$$\rho_1 u_1 \left(\frac{u_1^2}{2} + h_1\right) = \rho_2 u_2 \left(\frac{u_2^2}{2} + h_2\right), \tag{9.22}$$

where the subscripts 1 and 2 denote the values of the respective fields on the two sides of the jump. Using equation (9.20) we can simplify equation (9.22) to

$$\frac{u_1^2}{2} + h_1 = \frac{u_2^2}{2} + h_2, \tag{9.23}$$

which appears to be the conservation of Bernoulli's potential across the jump. (Note that the incompressible flow limit could be formally seen as the $\gamma \to \infty$ limit for which $h \to p/\rho$ and thus we recover Bernoulli's theorem (2.6).)

Relations (9.20), (9.21) and (9.23) are called the *Rankine-Hugoniot* jump conditions (in the frame of reference moving with the jump). One trivial solution to these conditions, in the polytropic gas case, is $u_1 = u_2 = 0$, $p_1 = p_2$ and arbitrary ρ_1 and ρ_2. This kind of jump is called *contact discontinuity* and it may arise due to a discontinuity in the initial state of the gas rather than as a result of wave breaking. The rest of the solutions are shocks. Since these are three conditions for six variables, $u_1, u_2, \rho_1, \rho_2, p_1$ and p_2, we can specify three of these variables, the choice of which depends on the particular physical problem. For example, for a shock moving into a still gas with a given state, one must specify ρ_1 and p_1 and either p_2 (if the shock is driven by a fixed increase in pressure or temperature) or the velocity downstream of the shock (if the shock is driven by a piston pushed in with a fixed speed); see question 9.3.7.

9.1.2 Two-dimensional gas dynamics

Two-dimensional dynamics of a compressible gas is a rather hard subject to study in general, and its general treatment is above the level at which this book is aimed. Here, we will only consider some simplified situations.

9.1.2.1 Irrotational isentropic flows

Vorticity and entropy are usually generated in the flows crossing 2D shocks, so our consideration here of the irrotational isentropic flows is relevant mostly to the situations where shocks are absent or very weak.

First of all one can derive a generalised version of Bernoulli's theorem for the irrotational isentropic flows which is expressed by the same equation (2.5) as for the incompressible flows, but the Bernoulli potential for the compressible flow is different:

$$\partial_t\phi + B = B_0 = \text{const}, \quad B = h + \frac{|\mathbf{u}|^2}{2}, \tag{9.24}$$

where ϕ is the velocity potential, $\mathbf{u} = \nabla\phi$ and the enthalpy h in the isentropic case is defined via the relation:

$$dh = \frac{dp}{\rho}. \tag{9.25}$$

In particular, for the polytropic gas we have $h = c_s^2/(\gamma-1)$; see the expression in (9.19). (The derivation of the compressible Bernoulli's law is nearly identical to the incompressible version, and we leave it to the reader.)

Further, for the polytropic gas, from the mass continuity equation we have:

$$\partial_t\rho = -\rho\nabla^2\phi - \nabla\rho \cdot \nabla\phi \quad \text{or} \quad \partial_t c_s^2 = -(\gamma-1)c_s^2\nabla^2\phi - \nabla c_s^2 \cdot \nabla\phi. \tag{9.26}$$

Substituting here for c_s^2 from Bernoulli's equation, in the polytropic gas case we have the following closed evolution equation for the velocity potential:

$$(\partial_t + \nabla\phi \cdot \nabla)\left(2\partial_t\phi + (\nabla\phi)^2\right) = (\gamma-1)\left(2B_0 - 2\partial_t\phi - (\nabla\phi)^2\right)\nabla^2\phi. \tag{9.27}$$

This is a nonlinear equation which is hard to solve in general. A significant simplification arises in the special case when the flow is almost uniform, e.g. arising when passing a slender body, $\mathbf{u} = U\hat{\mathbf{x}} + \nabla\tilde{\phi}$ with $U = \text{const}$ and $|\nabla\tilde{\phi}| \ll U$ as well as $|\nabla\tilde{\phi}| \ll \sqrt{B_0} \sim c_s$. In this case equation (9.27) can be linearised, and we have:

$$(\partial_t + U\partial_x)^2\tilde{\phi} = c_0^2\nabla^2\tilde{\phi}, \tag{9.28}$$

where c_0 is the speed of sound far from the body where the flow is uniform and homogeneous.

Of special importance is the case of a steady flow, $\partial_t \tilde{\phi} = 0$, in which case in place of equation (9.28) we have:

$$\left(M_0^2 - 1\right) \partial_{xx} \tilde{\phi} = \nabla_\perp^2 \tilde{\phi}, \qquad (9.29)$$

where $M_0 = U/c_0$ is the Mach number measured with respect to the undisturbed flow far from the body and $\nabla_\perp^2 = \partial_{yy} + \partial_{zz}$.

Equation (9.29) is very revealing: for subsonic flows, $M_0 < 1$, by the change $x' = x/\sqrt{1 - M_0^2}$ one can transform it to the Laplace equation, i.e. the same equation that describes irrotational flows in the incompressible case. Thus, modulo this change of variables, all results obtained for the incompressible case can be obtained for the compressible flow, as long as the linear approximation used above remains applicable. This also applies to the Zhukovskiy theorem; see question 9.3.9.

On the other hand, for supersonic flows, $M_0 > 1$, by the change $\tau = x/\sqrt{M_0^2 - 1}$ one can transform equation (9.29) to the wave equation, i.e. the same equation that describes linear sound waves. This fact can be effectively used for solving problems about supersonic flows around slender bodies: see questions 9.3.10 and 9.3.11.

One can also find a simplified version of Bernoulli's law for the linearised theory, which is useful for finding the pressure, which in turn is useful for calculating the drag and the lift forces. The steady-state version of equation (9.24) in the case of small disturbances is:

$$\tilde{p}/\rho_0 = -U\partial_x \tilde{\phi} - \frac{(\nabla_\perp \tilde{\phi})^2}{2}, \qquad (9.30)$$

where we have taken into account that, according the equation (9.25), $\tilde{h} = \tilde{p}/\rho_0 + O(\tilde{p}^2)$ (so the remainder is sub-dominant to either one of the terms in (9.30) or to both).

The linearised equation (9.29) is not valid for transonic flows, $|M_0 - 1| \ll 1$. Treatments of these cases are possible using the Euler-Tricomi equation, but we will omit this subject here (the interested reader is referred to the book of Landau and Lifshitz [14]). Also, the linearised equation (9.29) is not valid for very large Mach numbers: this case will be treated in the next section.

9.1.2.2 Steady hypersonic flow past a slender body

For very large Mach numbers the transverse velocity perturbation becomes of the same order as the speed of sound, $|\nabla_\perp \tilde{\phi}| \sim c_s$, and the linear approximation of the previous section is no longer valid. Luckily, a simplified description is available in this case too.

Let us consider a steady inviscid compressible 2D flow described by the steady-state version of equations (1.5) and (1.6):

$$u\partial_x u + v\partial_y u = -\frac{1}{\rho}\partial_x p, \tag{9.31}$$

$$u\partial_x v + v\partial_y v = -\frac{1}{\rho}\partial_y p, \tag{9.32}$$

$$\partial_x(\rho u) + \partial_y(\rho v) = 0. \tag{9.33}$$

Of course, the system of these equations is incomplete and we need to add another equation, e.g. the one for the entropy or the energy density. In view of using our system for describing shocks, like in 1D, it should be the energy rather than the entropy equation. However, we will not write it here explicitly because we will not need it for our discussion below.

Suppose that at infinity the flow has a constant uniform speed along the x-axis, $\mathbf{u} = U\hat{\mathbf{x}}$, $U = \text{const}$, and it passes a slender 2D body with a typical angle of inclination of its surface to the x-axis $\theta \ll 1$; see figure 9.1. Let the flow be hypersonic so that $M \geq 1/\theta$. Presence of slender a body produces small disturbances in the velocity field,

$$\mathbf{u} = (U + u')\hat{\mathbf{x}} + v'\hat{\mathbf{y}}$$

such that the leading order changes are in the transverse component only, $v' \sim \theta U \ll U$ and $u' \sim \theta^2 U \ll v'$. (One could say that the leading order change is in the direction of the velocity field rather than its absolute value.) On the other hand, these disturbances are not small compared to the speed of sound and, therefore, the linearised theory considered in the previous section is inapplicable.

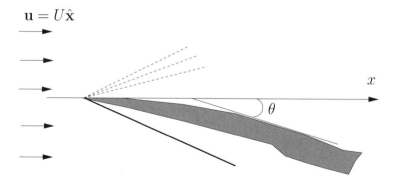

FIGURE 9.1: Hypersonic flow past a slender body ("hypersonic plane"). The bold solid line is a shock, dashed lines indicate an expansion fan.

The fact that the changes in the x-velocity are much weaker that the changes in the y-velocity allows us to neglect the former in equations (9.32) and (9.33) and write them as

$$\partial_\tau v + v\partial_y v = -\frac{1}{\rho}\partial_y p, \tag{9.34}$$

$$\partial_\tau \rho + \partial_y(\rho v) = 0, \tag{9.35}$$

where $\tau = x/U$ is an effective time. We have arrived at equations which are formally identical to equations (9.1) and (9.2) for an unsteady 1D flow. Without derivation we mention that the same equivalence also applies to the energy balance equation which we have omitted in this section.

Thus, solving a steady 2D problem of hypersonic flow about a slender body reduces to solving an unsteady 1D problem with an effective time $\tau = x/U$. In this analogy, the free-slip boundary condition on the solid surface $y = Y(x)$ becomes a boundary condition corresponding to a moving piston: $v = \partial_\tau Y(x(\tau))$ on $y = Y(x(\tau))$ where $x(\tau) = U\tau$. Because of this formal correspondence, the described approach is called the piston analogy. For example, in figure 9.1, the piston moves "into the gas" on the windward (bottom) part of the body and, therefore, a shock wave forms; see question 9.3.7. On the leeward (upper) side of the body the piston "withdraws from the gas" and, therefore, a rarefaction wave with its typical expansion fan of characteristics forms; see question 9.3.2.

Question 9.3.12 will exploit the piston analogy to find a lift force on a plane hypersonic wing.

9.2 Further reading

Detailed discussions of various aspects of the compressible gas dynamics can be found in the books *Fluid Dynamics* by L.D. Landau and E.M. Lifshitz [14], *Linear and Nonlinear Waves* by G.B. Whitham [31], "Waves in Fluids" by J. Lighthill [15] and *Prandtl's Essentials of Fluid Mechanics* by Oertel et al [17].

9.3 Problems

9.3.1 Characteristic equations

Derive the equations for a 1D gas in their characteristic form (9.4) and (9.5) with the characteristic curves defined by equations (9.7) and (9.8).

9.3.2   Flow due to a piston withdrawal

> Given information: the equations for a 1D gas in the characteristic form
> (9.4), (9.5) and (9.6) with the characteristic curves defined by equa-
> tions (9.7), (9.8) and (9.9). Characteristic equations for the isentropic
> gas (9.10) and (9.11) with the polytropic-gas expressions for the Rie-
> mann invariants (9.14).

If a flow is isentropic and one of the Riemann invariants is uniform in the physical space, the solution for it can be found analytically. This kind of flow is called the *simple wave*, and in the linear limit it corresponds to sound waves propagating in a single direction (i.e. either only to the right or only to the left). In the present problem we will encounter such a simple wave motion.

A piston suddenly starts moving on trajectory $x = X(t)$ withdrawing from a tube filled with gas which is initially still and has uniform initial pressure p_0 and density ρ_0. Our aim is to find the solution for the gas motion after the piston starts moving. (Note that the case where the piston is pushed into the gas is analysed in question 9.3.5.)

1. Consider the C^0-characteristics (i.e. the fluid paths) on the (x, t)-plane bounded by the piston's trajectory $x = X(t)$ placing the origin so that $X(0) = 0$. What entropy has the gas initially (i.e. on the x-axis at $x > 0$)? How does this value propagate along the C^0-characteristics? Will the C^0-characteristics originating on the x-axis cover the entire space occupied by the gas for any time $t > 0$? By answering these questions, find the entropy field for any time $t > 0$.

2. What is the initial value of the Riemann invariant R^-? Will the C^--characteristics originating on the x-axis cover the entire space occupied by the gas for $t > 0$? What is the initial value of the Riemann invariant R^-. Find R^- for all $t > 0$ and all $x > X(t)$.

3. Explain why the C^+-characteristics originating on the x-axis do not cover the entire space occupied by the gas at $t > 0$. What are the values of the Riemann invariant R^+ on the x-axis and on the piston's trajectory $x = X(t)$? Find R^+ for all $t > 0$ and all $x > X(t)$ considering separately the C^+-characteristics originating on the x-axis and on $x = X(t)$.

4. Put together the results of the previous parts and find an implicit so-lution for u and c_s in terms of a time parameter τ parametrising the piston's position (i.e. when $t = \tau$ the piston is at $x = X(\tau)$).

5. Find the explicit solution for u and c_s for the case when for $t > 0$ the piston has a constant speed $\dot{X} = -V < 0$. What is the limiting piston withdrawal speed above which a vacuum bubble forms?

9.3.3 Gas expansion into vacuum

A still gas with uniform pressure p_0 and density ρ_0 occupies the half-space $x > 0$ limited by a solid plate at $x = 0$. The second half-space, $x < 0$, is occupied by vacuum. At time $t = 0$ the solid plate is instantaneously removed. Find the gas motion for $t > 0$.

9.3.4 Dam break flow

Given information: the 1D shallow water equations:

$$\partial_t u + u \partial_x u = -g \partial_x h, \qquad (9.36)$$
$$\partial_t h + \partial_x (hu) = 0, \qquad (9.37)$$

where $u(x,t)$ and $h(x,t)$ are the fluid velocity and the surface height respectively and g is the gravity acceleration constant.

Still shallow water with height $h = h_0$ occupies the half-space $x > 0$ limited by a gate at $x = 0$. The second half-space, $x < 0$, is dry, $h = 0$. At time $t = 0$ the gate is instantaneously opened. Find the water motion for $t > 0$.

Hint: You may find it helpful to use the fact that the 1D shallow water equations are formally identical to the equations for 1D polytropic gases (9.1) and (9.2). What would be the adiabatic index γ in this case?

9.3.5 Momentum conservation in a viscous compressible flow

Compressible viscous 1D flow is described by the mass continuity equation (1.6) and by the Navier-Stokes equation

$$\rho(\partial_t u + u \partial_x u) = -\partial_x p + \lambda \partial_{yy} u, \quad \lambda = \xi + 4\mu/3, \qquad (9.38)$$

where μ and ξ are the first and the second viscosity coefficients respectively.

Derive the continuity equation for the momentum density in a 1D compressible viscous flow.

9.3.6 Energy conservation in compressible flow

1. Using the 1D compressible inviscid flow equations (9.1), (9.2), and (9.3) prove the energy continuity equation (9.17) for a 1D polytropic flow, in which the internal energy and the enthalpy are given by expressions (9.19).

2. Without derivation, present a general qualitative argument on the role of viscosity in the energy conservation law.

9.3.7 Rankine-Hugoniot conditions for jumps across shocks

1. Consider a 1D shock wave with a still gas ahead of it with pressure p_1 and density ρ_1. The pressure behind the shock is p_2. Using the Rankine-Hugoniot conditions (9.20), (9.21) and (9.23), and the polytropic gas expression for the enthalpy (9.19), find the gas density and velocity behind the shock, as well as the shock's speed.

2. Consider the limit of very strong shocks, $p_2/p_1 \to \infty$. Find the limiting density ratio ρ_2/ρ_1 for an ideal monatomic gas ($\gamma = 5/3$) and for an ideal diatomic gas ($\gamma = 1.4$). What limiting density ratio should we expect for air?

3. A piston is suddenly pushed with constant speed V into a tube filled with a gas which is initially still and which has initial pressure p_1 and density ρ_1. Describe the gas motion after the piston starts moving.

9.3.8 Hypersonic collision of two gas masses

Given information: jump conditions across the shocks are (c.f. question 9.3.7):

$$(u_1 - U_1)^2 = \frac{\rho_3\,(p_3 - p_1)}{\rho_1\,(\rho_3 - \rho_1)}, \tag{9.39}$$

$$\rho_3 = \rho_1 \frac{\frac{p_3}{p_1}(\gamma + 1) + (\gamma - 1)}{(\gamma + 1) + \frac{p_3}{p_1}(\gamma - 1)}, \tag{9.40}$$

where u_1, p_1 and ρ_1 are the velocity, the pressure and the density ahead of the shock, U_1 is the shock speed and p_3 and ρ_3 are the pressure and the density behind of the shock. (Here we have chosen the same notations as for the right shock in figure 9.2.)

Consider two polytropic gas clouds with initial uniform pressures and densities, p_1, ρ_1 and p_2, ρ_2, moving toward each other with velocities $u_1 < 0$ and $u_2 > 0$ respectively (gas 1 is on the right and gas 2 is on the left). At some instant of time the clouds collide, and our goal is to describe their subsequent dynamics assuming that the collision is hypersonic: the relative velocity $u = u_2 - u_1$ is much greater than the initial speed of sound in both clouds.

1. Explain why, in general, three jumps will form: two shocks running away from the collision point in the opposite directions an a contact discontinuity in between the shocks; see figure 9.2.

2. The shock waves that form as a result of a hypersonic collision are strong: the pressure behind the shock is much greater than the pressure ahead of the shock. Find the limiting expressions for the ratio of the density

behind the shock to the density ahead of the shock when the pressure
ratio tends to infinity (c.f. question 9.3.7). Find the densities on the two
sides of the contact discontinuity.

3. In the strong shock limit, using the mass flow continuity, find the shock
   speeds U_1 and U_2 in terms of the gas velocities u_1 and u_2 in the reference
   frame of stationary contact discontinuity.

4. Using (9.39), find the pressures behind the shocks in terms of the flow
   parameters ahead of the shocks and the shock speeds in the strong shock
   limit. Equating pressures on the two sides of the contact discontinuity,
   find another condition on the velocities and, using this condition, find
   U_1 and U_2 in terms of the relative velocity $u = u_2 - u_1$ in the reference
   frame of stationary contact discontinuity. Find the pressure between the
   shocks in terms of ρ_1, ρ_2, α and u.

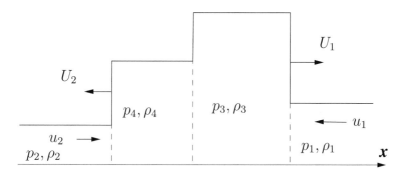

FIGURE 9.2: Gas motion after a collision of two clouds. The dashed lines
mark positions of the three jumps: two shocks (on the right and on the left)
and a contact discontinuity (in the middle). The solid line is the density profile.
The pressure profile is similar, except that there is no pressure jump across
the contact discontinuity.

9.3.9   Zhukovskiy's theorem for the subsonic flow

Given information:

- Zhukovskiy's theorem for the lift force F onto a wing in a steady inviscid incompressible 2D flow which is uniform at infinity, $\mathbf{u}(x, y) \to U\hat{\mathbf{x}}$ for $|\mathbf{x}| \to \infty$:

$$F = -\rho U\Gamma,$$

  where Γ is the velocity circulation around the wing (see equation (6.10)).

- The linearised equation for the velocity potential perturbation for the uniform compressible subsonic 2D flow with velocity $U\hat{\mathbf{x}}$ past a thin wing of uniform cross-section is (c.f. equation (9.29)):

$$\beta^2 \partial_{xx}\tilde{\phi} + \partial_{yy}\tilde{\phi} = 0, \tag{9.41}$$

  where $\beta = \sqrt{1 - M_0^2}$ and $M_0 = U/c_0$ is the Mach number measured with respect to the undisturbed flow far from the wing.

1. Formulate the free-slip boundary condition on the wing's surface and linearise this condition taking into account that both the deviation of the velocity field from the uniform field $U\hat{\mathbf{x}}$ and the deviation of the unit normal to the surface from $\hat{\mathbf{y}}$ are small.

2. Find a scaling transformation to reduce the equation (9.41) and the respective boundary condition to an effective incompressible flow problem.

3. Find the relation between the circulations and the flow velocities around the wing in the compressible flow and in the effective incompressible flow. By applying Zhukovskiy's lift theorem to the effective incompressible flow (assuming that the force in the compressible case will be the same as for the respective effective flow), find the lift force for the subsonic compressible flow over the thin wing.

9.3.10   Flow around a cone-nosed rocket

Given information:

- The linearised equation for the velocity potential perturbation for the uniform compressible supersonic flow with velocity $U\hat{\mathbf{x}}$ past a thin axisymmetric body is (c.f. the equation (9.29)):

$$\frac{1}{r}\partial_r(r\partial_r\tilde{\phi}) - \beta^2\partial_{xx}\tilde{\phi} = 0, \qquad (9.42)$$

  where $\beta = \sqrt{M_0^2 - 1}$ and $M_0 = U/c_0$ is the Mach number measured with respect to the undisturbed flow far from the wing.

- Bernoulli's law for small flow disturbances (9.30).

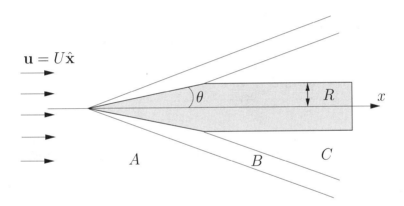

FIGURE 9.3: Flow past a thin rocket.

Consider a supersonic compressible flow with uniform velocity at infinity, $U\hat{\mathbf{x}}$ for $x, r \to \infty$, past a rocket with an axisymmetric shape: a conical nose and a cylindrical body; see figure 9.3. The rocket is thin and the linearised equation (9.42) is applicable. For simplicity we will consider the rocket to be infinitely long (spanning to $x \to \infty$ so that we can ignore the flow modifications by the tail part).

1. Identify the three different parts of the flow separated by the characteristics originating at the rocket's tip and at the circle separating the conical and the cylindrical parts of the body; see figure 9.3. What is the velocity field in the front and the back parts of the flow, i.e. the parts A and C in figure 9.3? Sketch the streamlines for the entire flow.

   What kind of discontinuities does the flow have on the surfaces separating the parts A and B and the parts B and C respectively in the linear approximation? In the more general case when the linear approximation is removed?

2. Consider the middle part of the flow (part B in figure 9.3). Seek a self-similar solution for the velocity potential perturbation in the form

$$\tilde{\phi}(r, x) = x f(\eta) \quad \text{with} \quad \eta = r/x.$$

By substituting this expression into equation (9.42), derive an ordinary differential equation (ODE) for the function $f(\eta)$.

3. Write down the free-slip boundary condition for the cone (nose) surface. Express this boundary condition in terms of a condition on the function $f(\eta)$.

4. Solve the ODE for the function $f(\eta)$ with the boundary condition and thereby find the flow for part B of the flow.

5. Using the small-disturbance Bernoulli law, find the pressure field on the rocket's surface and use the result to find the wave drag on the rocket.

9.3.11   Flow around a wedge

Given information:

- The linearised equation for the velocity potential perturbation for the uniform compressible supersonic 2D flow with velocity $U\hat{\mathbf{x}}$ past a thin wing of uniform cross-section is (c.f. equation (9.29)):

$$\beta^2 \partial_{xx}\tilde{\phi} - \partial_{yy}\tilde{\phi} = 0, \tag{9.43}$$

where $\beta = \sqrt{M_0^2 - 1}$ and $M_0 = U/c_0$ is the Mach number measured with respect to the undisturbed flow far from the wing.

- The Bernoulli law for small flow disturbances (9.30).

- The following relations hold for the shock created by a piston pushing with a constant velocity V into a 1D polytropic gas with density ρ_0 and pressure p_0 (see equations (9.69) and (9.70)):

$$p = p_0 + \rho_0 U_s V, \tag{9.44}$$

$$-\frac{U_s V}{2}(\gamma + 1) + U_s^2 - c_0^2 = 0, \tag{9.45}$$

where U_s is the shock speed, p is the pressure behind the shock and $c_0 = \sqrt{\gamma p_0/\rho_0}$ is the speed of sound ahead of the shock.

Consider a supersonic compressible 2D flow with uniform velocity $U\hat{\mathbf{x}}$ at

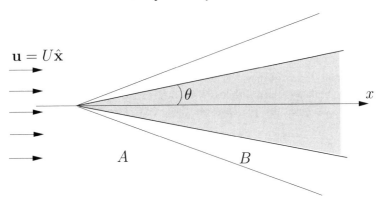

FIGURE 9.4: Flow past a wedge.

$x \to -\infty$ past a symmetric wedge with semi-angle θ and spanning from $x = 0$ to $x \to \infty$; see figure 9.4. The wedge is thin and the linearised equation (9.43) is applicable.

1. Identify the two different parts of the flow separated by the characteristics originating at the wedge's vertex; see figure 9.4. What is the velocity field in the front part of the flow, i.e. part A in figure 9.4? Sketch the streamlines for the entire flow.

   What kind of discontinuity does the flow have on the surfaces separating the parts A and B in the linear approximation? In the more general case when the linear approximation is removed?

2. Consider the back part of the flow—part B in figure 9.4. Find the solution for the velocity potential perturbation in this part by fitting the general solution of the 1D wave equation to the boundary condition arising from the free slip at the wedge's surface.

3. Using Bernoulli's law for small flow disturbances (9.30), find the pressure field on the wedge's surface.

4. Suppose now that $\beta\theta \sim 1$ and the linear approximation is not applicable. Formulate the piston analogy for the case $M \gg 1$ and thereby describe the flow field in this case, including the velocity, density and pressure in part B of the flow.

9.3.12 Lift force on a hypersonic wing

Given information:

- The following relations hold for the shock created by a piston pushing with a constant velocity V into a 1D polytropic gas with density ρ_0 and pressure p_0 (see equations (9.69) and (9.70)):

$$p = p_0 + \rho_0 U_s V, \qquad (9.46)$$

$$-\frac{U_s V}{2}(\gamma + 1) + U_s^2 - c_0^2 = 0, \qquad (9.47)$$

where U_s is the shock speed, p is the pressure behind the shock and $c_0 = \sqrt{\gamma p_0/\rho_0}$ is the speed of sound ahead of the shock.

- The following relation holds for the expansion wave created by a piston withdrawing with a constant velocity V from a 1D polytropic gas with density ρ_0 and pressure p_0 for the gas volume adjacent to the piston (to be more specific, for the gas volume between the piston and the expansion fan; see figure 9.7):

$$c_s = c_0 - \frac{\gamma - 1}{2}V \quad \text{for } V < \frac{2c_0}{\gamma - 1}, \quad \text{and } c_s = 0 \text{ otherwise,}$$
$$(9.48)$$

where again $c_0 = \sqrt{\gamma p_0/\rho_0}$; c.f. equation (9.57).

Consider a compressible 2D flow with uniform velocity $U\hat{\mathbf{x}}$ at $x \to -\infty$ past a flat wing placed under angle θ to the x-axis; see figure 9.5. The flow is hypersonic, $M = U/c_0 \gg 1$ and $M \sim 1/\theta$. The gas is polytropic with the adiabatic index γ.

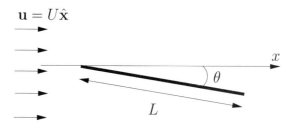

FIGURE 9.5: Hypersonic flow past a flat wing.

1. Describe the main features of the flow. Can the linearised theory be used? Formulate the piston analogy. Consider the flow discontinuities originating on both the windward and the leeward edges of the wing. What kind of discontinuities are they? Sketch the relevant shocks and

characteristics associated with these discontinuities. Sketch the stream-lines for the flow.

2. Using the piston analogy and the given information, find the pressure on both the upper and the lower parts of the wing. Use these results to find the lift force on the wing.

9.3.13 Formation of a blast wave by a very intense explosion

The name of this problem was exactly the title of two papers in 1950 by G.I. Taylor who used a simple formula to calculate the power of the world's first nuclear explosion which had previously been classified [25, 26]. Naturally, this revelation had a great impact on the military and its public relations.

A nuclear explosion in atmosphere of density ρ_0 deposits energy E and thereby produces a spherical blast shock wave whose distance from the explosion point r grows with time t. Assume that the explosion is so strong that the only dimensional parameters determining the shock motion are E, ρ_0 and t. In particular, the ambient atmospheric pressure is unimportant (c.f. question 9.3.7). Use dimensional analysis to find G.I. Taylor's relation between r, E, ρ_0 and t.

9.3.14 Balloon in polytropic atmosphere

A stratospheric balloon is partially filled with the buoyant gas hydrogen on the ground. As the balloon rises, it inflates and increases in volume. This leads to an additional lift. On the ground, the balloon has volume V_0, while its maximum volume is V_1. The atmospheric pressure and density at the ground level are p_0 and ρ_0 respectively. The hydrogen density at the ground level is ρ_{H_2}.

1. Assume that the state of the gas in the atmosphere changes polytropically,

$$p = p_0 \left(\frac{\rho}{\rho_0} \right)^{\gamma},$$

where γ is constant. Determine the dependence of the density of the atmosphere on the height z.

2. What is the greatest possible weight of the load G_{max} that can be lifted? (The balloon itself is part of the weight, but the buoyant gas is not).

3. What is the largest height to which weight G can be lifted? Assume that at this height the maximal volume V_1 is reached.

9.4 Solutions

9.4.1 Model solution to question 9.3.1

Differentiating the pressure and taking into account the equation (9.3), we have

$$\partial_t p + u \partial_x p = (\partial_\rho p)_S (\partial_t \rho + u \partial_x \rho) + (\partial_S p)_\rho (\partial_t S + u \partial_x S) = c_s^2 (\partial_t \rho + u \partial_x \rho). \tag{9.49}$$

Here, $(\partial_\rho \cdot)_S$ means the ρ-derivative with S kept constant and, respectively, $(\partial_S \cdot)_\rho$ means the S-derivative with ρ kept constant.

Multiplying the momentum equation (9.1) by $\pm \rho c_s$, continuity equation (9.2) by c_s^2 and adding the results to the pressure equation (9.49), we see that the derivatives of ρ cancel and we get

$$\pm \rho c_s \partial_t u \pm \rho c_s u \partial_x u + c_s^2 \rho \partial_x u + \partial_t p + u \partial_x p = \mp c_s \partial_x p. \tag{9.50}$$

By grouping the terms in this equation, we finally rewrite them as

$$[\partial_t + (u \pm c_s) \partial_x] p \pm \rho c_s [\partial_t + (u \pm c_s) \partial_x] u = 0. \tag{9.51}$$

These equations are the equations (9.4) and (9.5) with the characteristic curves defined by equations (9.7) and (9.8).

9.4.2 Model solution to question 9.3.2

1. The C^0-characteristics are the fluid paths. Let us first assume that the piston's withdrawal speed is not too fast for a vacuum bubble to form (the opposite case will be discussed in the end of the solution). Then one of the fluid paths belongs to the particle adjacent to the piston and, therefore, tracing its trajectory $x = X(t)$. The fluid particles at large distances from the piston are not affected by its motion: they are not moving i.e. their paths on the (x,t)-plane are straight vertical lines. In between these faraway paths and the piston the fluid paths continuously fill the whole space $x > X(t)$ for all $t > 0$; see figure 9.6. Initially, on the semi-line $t = 0$, $x > 0$, the entropy is uniform: $S = S_0 = \ln(p_0/\rho_0^\gamma)$. This value is propagated along the C^0-characteristics covering the entire space occupied by the gas for any time $t > 0$, i.e. $S = S_0$ for any $t > 0$ and all $x > X(t)$.

2. The initial value of the Riemann invariant R^- is $R_0^- = \frac{2c_0}{\gamma-1}$, where $c_0 = c_s(p_0, \rho_0) = \sqrt{\gamma p_0/\rho_0}$. The C^--characteristics move with velocities less (by c_s) than the fluid paths and, therefore, the C^--characteristics originating on the x-axis cover the entire space occupied by the gas for $t > 0$; see figure 9.6. Thus, the initial value of the Riemann invariant R^- will propagate over the entire gas and R^- for all $t > 0$ and all $x > X(t)$.

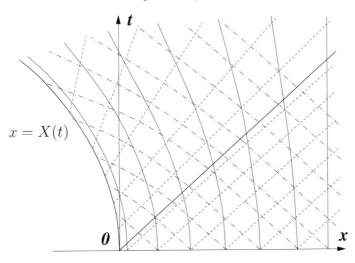

FIGURE 9.6: Characteristics in a piston-generated flow. C^0 – solid lines, C^- – dashed lines, C^+ – dash-dotted lines. The bold straight line passing through $(0,0)$ is separating the C^+-characteristics originating on the x-axis and the ones originating at the piston's trajectory $x = X(t)$.

3. The C^+-characteristics move with velocities greater (by c_s) than the fluid paths and, therefore, the C^+-characteristics originating on the x-axis do not cover the entire space occupied by the gas for $t > 0$, the solid line $x = c_0 t$ in figure 9.6 separates these characteristics from the ones that originate at the piston rather than the x-axis. The value of the Riemann invariant R^+ on the x-axis is $\frac{2c_0}{\gamma-1}$: it is propagated to the space below the solid line in figure 9.6. The value of the Riemann invariant R^+ on the piston's trajectory is $\frac{2c_s(\tau)}{\gamma-1} + \dot{X}(\tau)$ where τ parametrises the position on the piston's trajectory—it is the time at which the piston is at the position X. This value is propagated to the space above the solid line in figure 9.6 along characteristic $C^+(\tau)$ corresponding to $X(\tau)$.

4. Below the line $x = c_0 t$, since $R^- = R^+ = \frac{2c_0}{\gamma-1}$, we have $u = 0$ and $c_s = c_0$, i.e. the gas remains still and uniform, as in the initial state.

   Above the line $x = c_0 t$, we have:

$$R^- = \frac{2c_s}{\gamma - 1} - u = \frac{2c_0}{\gamma - 1} = \text{const} \quad - \text{ everywhere}, \qquad (9.52)$$

   and

$$R^+ = \frac{2c_s}{\gamma - 1} + u = \text{const} \quad \text{on} \quad C^+(\tau), \qquad (9.53)$$

   From these expressions we have:

$$u = \dot{X}(\tau) \quad \text{and} \quad c_s = c_0 + \frac{\gamma - 1}{2} u = c_0 + \frac{\gamma - 1}{2} \dot{X}(\tau), \qquad (9.54)$$

i.e. u and c_s are constant along the C_+-characteristics and these characteristics are are straight lines:

$$\dot{x}(t) = u + c_s = c_0 + \frac{\gamma + 1}{2} \dot{X}(\tau). \tag{9.55}$$

Integrating this equation with condition $x(\tau) = X(\tau)$, we have

$$x(t) = X(\tau) + \left[c_0 + \frac{\gamma + 1}{2} \dot{X}(\tau) \right] (t - \tau). \tag{9.56}$$

The implicit solution for u and c_s is given by (9.54) with the parameter $\tau = \tau(x, t)$ implicitly defined in (9.56).

5. For a constant speed $\dot{X} = -V < 0$, we have to distinguish three zones on the (x, t)-plane:

   (a) The zone of the C_+-characteristics originating on the x-axis at $x > 0$. This zone lies below the line $x = c_0 t$ and, as before, in this zone the gas remains still and uniform with the initial values of the pressure and the density.

   (b) The zone of the C_+-characteristics originating on the piston's trajectory at $t > 0$. This zone lies above the line $x = (c_0 - V)t$. Here from (9.54) we have:

$$u = -V \quad \text{and} \quad c_s = c_0 - \frac{\gamma - 1}{2} V. \tag{9.57}$$

   (c) The zone of the C_+-characteristics originating at $(0, 0)$. The characteristics in this zone are straight lines with all possible slopes between $1/c_0$ and $1/(c_0 - V)$. This is the so-called expansion fan; see figure 9.7. (Think of this zone as arising in a limiting case of gradual but vanishingly small acceleration of the piston from zero velocity to $-V$). Here $u + c_s = x/t$ and from (9.52) we have:

$$u = \frac{2c_0}{\gamma + 1} \left(\frac{x}{c_0 t} - 1 \right), \quad \text{and} \quad c_s = \frac{c_0}{\gamma + 1} \left((\gamma - 1) \frac{x}{c_0 t} + 2 \right).$$

Obviously, c_s cannot become negative. Thus, from (9.57) we have a critical speed of the piston,

$$V_* = \frac{2c_0}{\gamma - 1},$$

above which there is a vacuum region between the piston and the point $x = V_* t$.

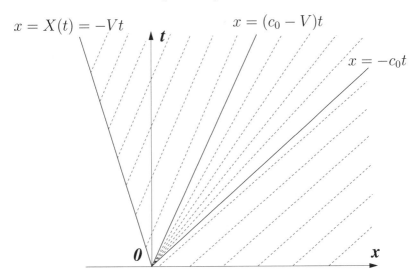

FIGURE 9.7: C^+-characteristics in a piston-generated flow in the case when the piston's speed is constant and negative, $\dot{X}(t) = -V < 0$.

9.4.3   Model solution to question 9.3.3

The solution to this problem is identical to the one given for the piston withdrawal problem in the case when the piston's speed is supercritical, $V > V_* = \frac{2c_0}{\gamma - 1}$; see question 9.3.2 and the respective solution 9.4.2. Indeed, in this case a vacuum bubble forms between the piston and the front of the rarefaction wave, i.e. the piston's motion is completely detached from the gas expansion, which occurs as if there was no piston at all.

9.4.4   Model solution to question 9.3.4

After replacing $h \to \rho$, the 1D shallow water equations (9.36) and (9.37) are formally identical to the equations for the 1D polytropic gas with $\gamma = 2$ in the isentropic case; see (9.1) and (9.2).

Thus, the present problem is just a version of question 9.3.3 with $\gamma = 2$.

9.4.5   Model solution to question 9.3.5

Multiplying equation (1.6) by u and adding the result to (9.38), we have

$$\partial_t(\rho u) + \partial_x(\rho u^2 + p - \lambda \partial_x u) = 0. \qquad (9.58)$$

which is the desired continuity equation for the momentum density in a 1D compressible viscous flow.

Note that for small viscosity the momentum flux modifications are large within the jump (shock) regions only, and, therefore, the inviscid version of the momentum equation (equation (9.15)) can be used for finding the jump conditions.

9.4.6 Model solution to question 9.3.6

1. Substituting expressions (9.19) into equation (9.17) and rearranging the terms we obtain:

$$\partial_t \left(\frac{p}{\gamma - 1} + \frac{\rho}{2} u^2 \right) = -\partial_x \left[\rho \frac{u^3}{2} + u \frac{\gamma p}{(\gamma - 1)} \right]. \tag{9.59}$$

From the equations (9.2) and (9.15) we have

$$\partial_t(\rho u^2) = \rho u \partial_t u + u \partial_t(\rho u) = -\rho u^2 \partial_x u - u \partial_x p$$
$$-u \partial_x(\rho u^2 + p) = -\partial_x(\rho u^3) - 2u \partial_x p. \tag{9.60}$$

On the other hand, using (9.2) and (9.3) we have

$$\partial_t p = (\partial_\rho p)_S \partial_t \rho + (\partial_S p)_\rho \partial_t S = -(\partial_\rho p)_S \partial_x(\rho u) - (\partial_S p)_\rho u \partial_x S \tag{9.61}$$

and

$$\partial_x p = (\partial_\rho p)_S \partial_x \rho + (\partial_S p)_\rho \partial_x S. \tag{9.62}$$

Here, $(\partial_S p)_\rho$ and $(\partial_\rho p)_S$ denote the derivative of p with respect to ρ with S fixed and the derivative of p with respect to S with ρ fixed. Combining the last two equations and using $(\partial_\rho p)_S = \gamma p / \rho$ we have

$$\partial_t p + u \partial_x p = -(\partial_\rho p)_S \partial_x(\rho u) - (\partial_S p)_\rho u \partial_x S + u (\partial_\rho p)_S \partial_x \rho +$$
$$u (\partial_S p)_\rho \partial_x S = -\rho (\partial_\rho p)_S \partial_x u = -\gamma p \partial_x u. \tag{9.63}$$

Substituting this and (9.60) into (9.59) we have

$$-\gamma \frac{p \partial_x u}{\gamma - 1} - u \partial_x p = -\partial_x(pu) - \frac{p \partial_x u}{(\gamma - 1)} = 0, \tag{9.64}$$

from which it is clear that the equality is identically satisfied, i.e. the energy continuity equation (9.17) is valid.

2. The energy conservation follows from the first principles of physics and as such it is the most fundamental in fluid dynamics, with or without dissipation, and even if additional processes are involved: thermoconductivity, ionisation, electromagnetic interactions. Of course, adding such additional effects would modify the expression for the internal energy and could also lead to further contributions to the energy expression (e.g. electro-magnetic energy, energy of interaction of the fluid with an external field). The role of viscosity is to transfer the kinetic equation into the internal one while leaving the total energy unchanged. On the other hand, such increase of the internal energy leads to an increase of the entropy.

9.4.7 Model solution to question 9.3.7

1. From equations (9.20) and (9.21) we get for u_1 and u_2:

$$u_1^2 = \frac{\rho_2}{\rho_1} \frac{(p_2 - p_1)}{(\rho_2 - \rho_1)}, \qquad (9.65)$$

$$u_2^2 = \frac{\rho_1}{\rho_2} \frac{(p_2 - p_1)}{(\rho_2 - \rho_1)}. \qquad (9.66)$$

Substituting these expressions into equation (9.23) yields the following relationship:

$$2\,(h_2 - h_1) = (p_2 - p_1)\left(\frac{1}{\rho_1} + \frac{1}{\rho_2}\right).$$

Using the polytropic gas expression for the entalpy (9.19), we get

$$\rho_2 = \rho_1 \frac{\frac{p_2}{p_1}(\gamma + 1) + (\gamma - 1)}{(\gamma + 1) + \frac{p_2}{p_1}(\gamma - 1)}. \qquad (9.67)$$

Thus, given ρ_1, p_1 and p_2, one can find ρ_2 from (9.67) and then find u_1 and u_2 using (9.65) and (9.66) respectively. After that we move to the frame where the gas ahead of the shock is stationary, by shifting the velocity field by $-u_1$. Thus in this frame the shock's speed will be $-u_1$ and the downstream gas velocity will be $u_2 - u_1$ (where u_2 and u_1 are still the velocities measured in the frame of the stationary shock).

2. As the shock's strength increases, there is an increase in the downstream gas pressure p_2, but the density ratio ρ_2/ρ_1 approaches a finite limit. From equation (9.67) we get for the upper limit on the density ratio:

$$\rho_2/\rho_1 = (\gamma + 1)/(\gamma - 1). \qquad (9.68)$$

It is equal to 4 for an ideal monatomic gas ($\gamma = 5/3$) and equal to 6 for an ideal diatomic gas ($\gamma = 1.4$). Since air is comprised mostly of diatomic molecules (nitrogen and oxygen), the limiting density ratio for shocks in air is approximately equal to 6.

3. The piston's motion will create a shock wave. In this case, the piston's velocity will be equal the gas velocity downstream of the shock, i.e. $V = u_2 - u_1$, whereas before u_1 and u_2 are measured in the stationary shock frame. In the laboratory frame, the shock's speed is $U = -u_1$.

From equations (9.20) and (9.21) we have:

$$p_2 = \rho_1 u_1^2 + p_1 - \rho_1 u_1 u_2 = p_1 + \rho_1 UV. \qquad (9.69)$$

Using this in equation (9.23) yields:

$$\frac{u_1^2}{2} + \frac{\gamma p_1}{(\gamma - 1)\rho_1} = \frac{u_2^2}{2} + \frac{\gamma p_2}{(\gamma - 1)\rho_2} = \frac{(V - U)^2}{2} - \frac{\gamma(p_1 + \rho_1 UV)(V - U)}{(\gamma - 1)\rho_1 U}.$$

Cancelling on both sides $\frac{u_1^2}{2} = \frac{U^2}{2}$ and $\frac{\gamma p_1}{(\gamma-1)\rho_1}$ and multiplying the resulting equation by $(\gamma - 1)U/V$, we have

$$-\frac{UV}{2}(\gamma + 1) + U^2 - c_1^2 = 0. \qquad (9.70)$$

The positive solution of this quadratic equation for U gives the shock speed. Substituting U into (9.69) gives the pressure behind the shock.

9.4.8   Model solution to question 9.3.8

1. In general, it would be impossible to find a two-shock solution without a contact discontinuity which would satisfy the boundary conditions, i.e. matching to the velocities, pressures and densities of the flows at $x \to \pm\infty$: the system would be overdefined as it would contain one more equation to satisfy compared to the number of available variables. Hence in general there must be three jumps: two shocks running away from the collision point in the opposite directions and a contact discontinuity in between the shocks as in figure 9.2. The contact discontinuity removes one equation that has to be satisfied, namely the condition that the densities behind the two shocks must coincide.

2. According to equation (9.40), the limiting density ratio realised in strong shocks, when the pressure ratio tends to infinity is

$$\alpha = \frac{\gamma + 1}{\gamma - 1}$$

(c.f. question (9.3.7)). Thus, for the densities on the two sides of the contact discontinuity we have

$$\rho_3 = \alpha\rho_1 \quad \text{and} \quad \rho_4 = \alpha\rho_2.$$

3. The mass flow continuity across the two shocks yields respectively:

$$\rho_3 U_1 = \rho_1(U_1 - u_1) \quad \text{and} \quad \rho_4 U_2 = \rho_2(U_2 - u_2),$$

where the velocities are measured in the reference frame of the stationary contact discontinuity, $u_3 = u_4 = 0$.

Thus in the strong shock limit we have:

$$U_1 = -\frac{u_1}{\alpha - 1} \quad \text{and} \quad U_2 = -\frac{u_2}{\alpha - 1} \qquad (9.71)$$

(in the reference frame of the stationary contact discontinuity).

4. Using (9.39), we have for the pressures behind the shocks in the strong

shock limit (i.e. neglecting p_1 and p_2 compared to p_3 an p_4 and taking the limiting density ratio α):

$$p_3 = \frac{\alpha - 1}{\alpha} \rho_1 (U_1 - u_1)^2 \quad \text{and} \quad p_4 = \frac{\alpha - 1}{\alpha} \rho_2 (U_2 - u_2)^2. \quad (9.72)$$

From these expressions, equating pressures on the two sides of the contact discontinuity, $p_3 = p_4$, we find:

$$\rho_1 (U_1 - u_1)^2 = \rho_2 (U_2 - u_2)^2.$$

Substituting $u = u_2 - u_1$ and using the equations (9.71) we get:

$$u_1 = -\frac{u}{1 + \sqrt{\rho_1/\rho_2}} \quad \text{and} \quad u_2 = \frac{u}{1 + \sqrt{\rho_2/\rho_1}}, \quad (9.73)$$

so that

$$U_1 = \frac{u}{(\alpha - 1)(1 + \sqrt{\rho_1/\rho_2})} \quad \text{and} \quad U_2 = -\frac{u}{(\alpha - 1)(1 + \sqrt{\rho_2/\rho_1})}. \quad (9.74)$$

The pressure between the shocks can be found from (9.72):

$$p_3 = p_4 = \frac{\alpha u^2}{(\alpha - 1)(\sqrt{1/\rho_1} + \sqrt{1/\rho_2})^2}. \quad (9.75)$$

9.4.9   Model solution to question 9.3.9

1. The free-slip boundary condition on the wing surface $y = Y_{\pm}(x)$ says that the normal to the boundary component of velocity is zero (signs plus and minus correspond to the top and the bottom of the wing respectively). This is the same as to say that the vector product of the velocity and the tangential vector is zero. Since the ratio of the y- and the x-components of the tangential vector is $\partial_x Y(x)$, we have:

$$(U + \partial_x \tilde{\phi}) \partial_x Y_{\pm} - \partial_y \tilde{\phi} = 0 \quad \text{on} \quad y = Y_{\pm}(x).$$

Linearising this condition taking into account that $|\partial_x \tilde{\phi}| \ll U$ and $\partial_x Y_{\pm}(x) \ll 1$, we have:

$$U \partial_x Y_{\pm} - \partial_y \tilde{\phi} = 0 \quad \text{on} \quad y = Y_{\pm}(x). \quad (9.76)$$

2. By using a scaling transformation $\tilde{\phi}(x) = \phi'(x')/\beta$ and $Y_{\pm}(x) = Y'_{\pm}(x')$ with $x' = x/\beta$, we reduce equation (9.41) and the boundary condition (9.76) to an effective incompressible flow problem:

$$\partial_{x'x'} \phi' + \partial_{yy} \phi' = 0, \quad (9.77)$$

(this is the 2D Laplace equation) and

$$U \partial_{x'} Y'_{\pm} - \partial_y \phi' = 0 \quad \text{on} \quad y = Y'_{\pm}(x'). \quad (9.78)$$

3. For the circulation we have:

$$\Gamma' = \oint_{C'} \tilde{\mathbf{u}}(x', y) \cdot d\mathbf{s}' \approx \int [u'(x', y^-(x')) + v'(x', y^-(x'))\partial_{x'} Y'_-(x') -$$

$$u'(x', y^+(x')) + v'(x', y^+(x'))\partial_{x'} Y'_+(x')]dx'$$

$$= \beta^{-1} \int [u(x, y^-(x)) + v(x, y^-(x))\partial_x Y_-(x) -$$

$$u(x, y^+(x)) + v(x, y^+(x))\partial_x Y_+(x)]dx = \beta^{-1}\Gamma.$$

4. By applying the Zhukovskiy lift theorem to the effective incompressible flow, we find the lift force for the subsonic compressible flow over the thin wing:

$$F = -U\rho\Gamma/\beta.$$

Note that this value is greater than the lift force in an incompressible fluid—it is easier to fly in a compressible gas!

9.4.10   Model solution to question 9.3.10

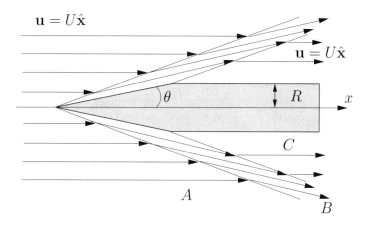

FIGURE 9.8: Flow past a thin rocket.

1. The characteristics lie at angle $\arctan(1/\beta)$ to the symmetry axis (Mach angle); they form conical surfaces. The flow ahead of the first characteristic cone (in part A in figure 9.8) is undisturbed, $\tilde{\phi} \equiv 0$ and $\mathbf{u} = U\hat{\mathbf{x}}$. The flow behind the last characteristic cone (in part C in figure 9.8) is calmed and the velocity field is the same as in part A, $\tilde{\phi} \equiv 0$ and $\mathbf{u} = U\hat{\mathbf{x}}$. Thus the streamlines for the entire flow are as shown in figure 9.8).

The 3D stationary linearised problem is equivalent to a 2D linear wave propagation problem with effective time $\tau = x/\beta$. The changing cross-section of the body then plays the role of a moving piston. Since the velocity of such a piston changes suddenly at the vertex of the cone (tip of the rocket) and the cone's base (boundary between the conical nose and the cylindrical body), the velocity field will have discontinuous gradients at the respective characteristic cones, i.e. on the surfaces separating parts A and B and parts B and C respectively. In the more general case, when the linear approximation is removed, the front discontinuity becomes a (weak) shock wave and the trailing discontinuity becomes a (weak) rarefaction wave (with an expansion fan concentrated in the narrow sector of angles).

2. For the middle part of the flow (part B in figure 9.8) we seek a self-similar solution for the velocity potential perturbation in the form $\tilde{\phi}(r, x) = xf(\eta)$ with $\eta = r/x$. By substituting this expression into equation (9.42), we get the following ODE for the function $f(\eta)$,

$$\eta(1 - \beta^2\eta^2)\partial_{\eta\eta}f + \partial_\eta f = 0. \tag{9.79}$$

3. The free-slip boundary condition for the cone surface is

$$u_\perp/(U + u_\parallel) \approx u_\perp/U = \tan\theta \approx \theta \quad \text{on} \quad r = x\tan\theta \approx x\theta. \tag{9.80}$$

This boundary condition in terms of a condition on the function $f(\eta)$ reads:

$$\partial_\eta f = U\theta \quad \text{at} \quad \eta = \theta. \tag{9.81}$$

4. The ODE has a trivial solution $f = \text{const}$ which corresponds to $\phi = \text{const}\, x$ and which can be discarded because it corresponds to a finite uniform x-velocity perturbation at infinity (the latter must be zero). The second fundamental solution is found by solving equation (9.79) with respect to $\partial_\eta f$, which, taking into account the boundary condition (9.81), gives:

$$\partial_\eta f = \partial_r\tilde{\psi} = u_r = \frac{\theta^2\beta U}{\sqrt{1 - \beta^2\theta^2}}\sqrt{\frac{1}{\beta^2\eta^2} - 1} = \theta^2\beta U\sqrt{\frac{x^2}{\beta^2 r^2} - 1}. \tag{9.82}$$

Here we took into account that $\beta\theta \ll 1$ because only for such values the linear approach we are using is valid; see section 9.1.2.2. Obviously, expression (9.82) is valid only for $r < x/\beta$ (recall that for $r > x/\beta$ we have part A of the flow where the velocity disturbance is zero).

Integrating equation (9.82) gives:

$$f = \theta^2 U\left[\sqrt{1 - \beta^2\eta^2} - \text{arccosh}\frac{1}{\beta\eta}\right]. \tag{9.83}$$

For the x-velocity disturbance we have:

$$\tilde{u}_x = \partial_x\tilde{\phi} = f - \eta\partial_\eta f = -\theta^2 U\,\text{arccosh}\frac{x}{\beta r}. \tag{9.84}$$

5. From the weak-disturbance version of Bernoulli's law we have

$$\tilde{p}/\rho_0 = -U\tilde{u}_x - \frac{u_r^2}{2} = \theta^2 U^2 \mathrm{arccosh}\frac{x}{\beta r} - \frac{\theta^4 U^2}{2}(x^2/r^2 - \beta^2). \quad (9.85)$$

On the cone surface $r = \theta x \ll x/\beta$ we have

$$\tilde{p}/\rho_0 = \theta^2 U^2 \mathrm{arccosh}\frac{1}{\beta\theta} - \frac{\theta^2 U^2}{2} \approx \theta^2 U^2 \ln\frac{2}{\beta\theta}, \quad (9.86)$$

where we have taken into account that for large z: $\mathrm{arccosh}z \approx \ln 2z \gg 1$.

The wave drag on the rocket D is determined by the net x-component of the pressure force. Since the pressure is the same at any point on the cone,

$$D = \pi R^2 \tilde{p} = \pi R^2 \theta^2 \rho U^2 \ln\frac{2}{\beta\theta}. \quad (9.87)$$

Note that we took here $\tilde{p} = p - p_0$ assuming that there is pressure p_0 at the tail part of the rocket (which we have not considered here).

9.4.11 Model solution to question 9.3.11

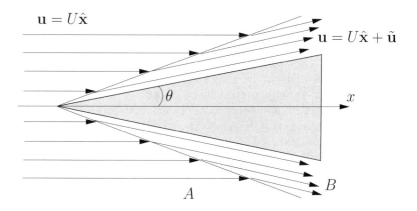

FIGURE 9.9: Flow past a wedge.

1. The characteristics originating at the vertex lie at angle $\arctan(1/\beta)$ to the x-axis (Mach angle). The flow ahead of these characteristics (in part A in figure 9.9) is undisturbed, $\tilde{\phi} \equiv 0$ and $\mathbf{u} = U\hat{\mathbf{x}}$. The streamlines for the entire flow are shown in figure 9.9). The streamlines change the direction as they enter into part B: they have to go parallel to the wedge's surface to satisfy the free-slip condition.

2. The flow behind these characteristics (in part B in figure 9.9) can be

found by solving equation (9.43) which is equivalent to the 1D nonstationary wave equation with an effective time $\tau = x/\beta$. The changing thickness of the body then plays the role of a moving piston. Since the velocity of such a piston changes suddenly at the vertex of the cone, the velocity field will have a discontinuous gradient at the characteristics originating at the wedge's vertex, i.e. on the interface separating parts A and B. In the more general case, when the linear approximation is removed, the front discontinuity becomes a (weak) shock wave.

3. The general solution of equation (9.43), like for the nonstationary 1D wave equation, consists of two counter-propagating waves:

$$\tilde{\phi} = f(y - x/\beta) + g(y + x/\beta), \qquad (9.88)$$

where f and g are arbitrary functions which have to be fixed using boundary conditions.

Because $\mathbf{u} \to U\hat{\mathbf{x}}$ at $x \to -\infty$, we have $g \equiv 0$ above the wedge and $f \equiv 0$ below the wedge.

The free-slip boundary condition for the cone surface is

$$\tilde{u}_y/(U + \tilde{u}_x) \approx \tilde{u}_y/U = \tan\theta \approx \theta \quad \text{on} \quad y = \pm x\tan\theta \approx \pm x\theta, \quad (9.89)$$

where the signs plus and minus correspond to the upper and the lower surfaces of the wedge respectively. This gives $\partial_y f = -\partial_y g = U\theta$ i.e. $f = -g = \pm U\theta(y \mp x/\beta)$.

Thus, in the linear approximation we have:

$$\tilde{u}_x = -U\theta/\beta \quad \text{and} \quad \tilde{u}_y = \pm U\theta, \qquad (9.90)$$

where the signs plus and minus correspond to the upper and the lower surfaces of the wedge respectively. Obviously, this expression is valid only for $y < x/\beta$ (recall that for for $y > x/\beta$ we have part A of the flow where the velocity disturbance is zero). Thus, the velocity in part B is constant and parallel to the respective surfaces of the wedge. Correspondingly, the streamlines are also parallel to the respective surfaces of the wedge; see figure 9.9).

4. Using the Bernoulli law (9.30), we have:

$$\tilde{p} = -U\tilde{u}_x - \tilde{u}_y^2/2 = U^2\theta/\beta - U^2\theta^2/2 \approx U^2\theta/\beta. \qquad (9.91)$$

5. If $\beta\theta \sim 1$, the transverse velocity is of the order of the speed of sound and the linear approximation is not applicable: a strong shock wave forms. However, if $M \gg 1$ (where M is measured with respect to the flow at $x = -\infty$, i.e. $M = U/c_0$) the 2D steady gas motion is equivalent to a 1D unsteady gas motion with an effective time $\tau = x/U$—this is

the piston analogy. Thus the flow is equivalent to the 1D gas motion caused by a piston which suddenly starts pushing into the gas with a constant velocity $V = \pm U \tan \theta \approx \pm U\theta$ (plus for the upper and minus for the lower sides of the wedge). The shock speed U_s in the 1D problem is related to angle α at which the 2D shock is oriented with respect to the x-axis via $\tan \alpha = U_s/U$. Part B of the flow corresponds to the flow behind such a 1D shock, and its parameters can be found from solving equations (9.44) and (9.45), which gives:

$$\tan \alpha = \pm(\gamma + 1)\theta/4 \pm \sqrt{\theta^2(\gamma + 1)^2/16 + 1/M^2} \qquad (9.92)$$

and

$$p = p_0 + \rho_0 U^2 \theta \left(\theta(\gamma + 1)/4 + \sqrt{\theta^2(\gamma + 1)^2/16 + 1/M^2} \right). \qquad (9.93)$$

Obviously, the y-velocity in part B is still as in (9.90) (i.e. in the respective 1D problem the velocity behind the shock is the same as the piston's velocity).

Note that expression (9.93) agrees with (9.91) in the limit $1 \ll M \ll 1/\theta$, i.e. when the applicability ranges of both approaches overlap.

9.4.12   Model solution to question 9.3.12

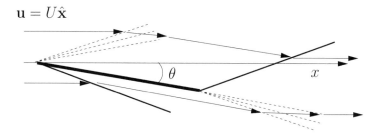

FIGURE 9.10: Hypersonic flow past a flat wing. The wing is shown by the very bold line. The bold line marks the shocks, and the dashed lines mark the expansion fans. The streamlines are marked by the thin solid lines.

1. When $M\theta \sim 1$ the transverse velocity is of the order of the speed of sound and the linear approximation is not applicable: strong shocks and nonlinear rarefaction waves form. However, if $M \gg 1$, the 2D steady gas motion is equivalent to a 1D unsteady gas motion with an effective time $\tau = x/U$—this is the piston analogy. Thus the flow below the wing is equivalent to the 1D gas motion caused by a piston which suddenly starts pushing into the gas with a constant velocity $V = -U \tan \theta \approx -U\theta$, whereas the flow above the wing is equivalent to the 1D gas motion

caused by a piston which suddenly starts withdrawing from the gas with a constant velocity $V = -U \tan \theta \approx -U\theta$. Respectively, the flow forms a shock discontinuity originating at the windward edge of the wing and extending below it, and an expansion fan of characteristics also originating at the windward edge and extending above the wing (there are velocity gradient discontinuities at the both boundaries of the expansion fan). At the leeward edge, the picture of the discontinuities is symmetric with respect to the windward edge, and behind downwind from these discontinuities the flow restores the original shape, $\mathbf{u} = U\hat{\mathbf{x}}$; see figure 9.10. This figure also shows a sketch of the streamlines.

2. By the piston analogy, for the flow below the wing we can use the result for the flow behind a 1D shock, equations (9.46) and (9.47). The shock speed U_s in the 1D problem is related to angle $\alpha < 0$ at which the 2D shock is oriented with respect to the x-axis via $\tan \alpha = U_s/U$. Solving (9.47) for U_s and substituting the result into (9.46), we have for the pressure below the wing:

$$p_- = p_0 + \rho_0 U^2 \theta \left(\theta(\gamma+1)/4 + \sqrt{\theta^2(\gamma+1)^2/16 + 1/M^2} \right); \quad (9.94)$$

c.f. equation (9.93).

Also by the piston analogy, for the flow above the wing we can use the result for the flow the 1D expansion wave created by a piston withdrawing with a constant velocity $V = U\theta$ from a 1D polytropic gas, namely equation (9.48). Taking into account that the rarefaction wave flow is isentropic, $c_s/c_0 = (p/p_0)^{(\gamma-1)/2\gamma}$, we have for the pressure above the wing:

$$p_+ = p_0 \left[1 - \frac{\gamma-1}{2} M\theta \right]^{2\gamma/(\gamma-1)} \quad \text{for} \quad M\theta < \frac{2}{\gamma-1}, \quad (9.95)$$

and $p = 0$ otherwise (vacuum).

The lift force on the wing (per unit length of the wing) is

$$F = (p_- - p_+)L\cos\theta \approx (p_- - p_+)L.$$

9.4.13   Model solution to question 9.3.13

The unit of energy is erg which is $g\,cm^2/s^2$ (just think of the kinetic energy expression $E = mu^2/2$). The unit of density ρ is g/cm^3 and the units of r and t are cm and s respectively. The only dimensional combination of E, ρ_0 and t that gives the correct dimension of r is:

$$r(t) = C \frac{\rho_0^{1/5}}{E^{1/5}} t^{2/5},$$

where C is a dimensionless constant (in polytropic gases it depends on the adiabatic index γ). This is G.I. Taylor's relation [26].

9.4.14   Model solution to question 9.3.14

1. The vertical component of the Euler equation is:

$$\frac{1}{\rho}\frac{\partial p}{\partial z} = -g.$$

Substituting the polytropic law, we have:

$$\frac{\gamma p_0 \rho^{\gamma-2}}{\rho_0^\gamma}\frac{\partial \rho}{\partial z} = -g.$$

Integrating this, we find:

$$\frac{\gamma}{\gamma-1}\frac{p_0}{\rho_0^\gamma}\rho^{\gamma-1} = -gz + C,$$

and matching to the boundary condition at the ground, $\rho(0) = \rho_0$,

$$\rho = \rho_0(1 - z/H)^{1/(\gamma-1)}, \tag{9.96}$$

where $H = \frac{p_0\gamma}{\rho_0 g(\gamma-1)}$.

2. Balancing the load weight and the buoyancy force at the ground level, we have:

$$G_{max} = gV_0(\rho_0 - \rho_{H_2}).$$

3. At height z_{max}, from (9.96) we have the following air density:

$$\rho_1 = \rho_0(1 - z_{max}/H)^{1/(\gamma-1)}. \tag{9.97}$$

So,

$$z_{max} = H - H(\rho_1/\rho_0)^{(\gamma-1)}. \tag{9.98}$$

The force balance at z_{max} is:

$$G = gV_1(\rho_1 - \rho'_{H_2}), \tag{9.99}$$

where ρ'_{H_2} is the hydrogen density at height z_{max}, which can be obtained from the mass conservation $\rho'_{H_2} = V_0\rho_{H_2}/V_1$. So, $\rho_1 = V_0\rho_{H_2}/V_1 + G/gV_1$.

Finally,

$$z_{max} = H - H\left[\frac{V_0\rho_{H_2} + G/g}{V_1\rho_0}\right]^{(\gamma-1)}.$$

Bibliography

[1] D.J. Acheson. *Elementary Fluid Dynamics*. Oxford Applied Mathematics and Computing Science Series. Clarendon Press, 1990.

[2] Hassan Aref, Paul K. Newton, Mark A. Stremler, Tadashi Tokieda, and Dmitri L. Vainchtein. Vortex Crystals. In *Advances in Applied Mechanics*, volume 39, pages 1–79. Elsevier, 2003.

[3] G.K. Batchelor. *The Theory of Homogeneous Turbulence*. Cambridge Science Classics. Cambridge University Press, 1953.

[4] G.K. Batchelor. *An Introduction to Fluid Dynamics*. Cambridge Mathematical Library. Cambridge University Press, 2000.

[5] P.G. Drazin and W.H. Reid. *Hydrodynamic Stability*. Cambridge Mathematical Library. Cambridge University Press, 2004.

[6] G. Falkovich. *Fluid Mechanics: A Short Course for Physicists*. Cambridge University Press, 2011.

[7] R. Fjørtoft. On the changes in the spectral distribution of kinetic energy for two-dimensional nondivergent flow. *Tellus*, 5(5):225, 1953.

[8] U. Frisch. *Turbulence: The Legacy of A. N. Kolmogorov*. Cambridge University Press, 1995.

[9] T. Kambe. *Elementary Fluid Mechanics*. World Scientific, 2007.

[10] A.N. Kolmogorov. The local structure of turbulence in incompressible viscous fluid for very large Reynolds numbers. *Proc. R. Soc. Lond. A*, 434:9–13, 1991.

[11] R.H. Kraichnan and D. Montgomery. Two-dimensional turbulence. *Reports on Progress in Physics*, 43(5):547, 1980.

[12] P.K. Kundu and I.M. Cohen. *Fluid Mechanics*. Elsevier Science, 2010.

[13] H. Lamb. *Hydrodynamics*. Dover Publications, 1945.

[14] L.D. Landau and E.M. Lifshitz. *Fluid Dynamics*. Theoretical Physics. Butterworth-Heinemann, 1995.

[15] J. Lighthill. *Waves in Fluids*. Cambridge Mathematical Library. Cambridge University Press, 2001.

[16] J.C. McWilliams. *Fundamentals of Geophysical Fluid Dynamics*. Cambridge University Press, 2006.

[17] H. Oertel, P. Erhard, K. Asfaw, D. Etling, U. Muller, U. Riedel, K.R. Sreenivasan, and J. Warnatz. *Prandtl's Essentials of Fluid Mechanics*. Applied Mathematical Sciences. Springer, 2010.

[18] L. Onsager. Statistical hydrodynamics. *Nuovo Cimento*, 6(2):279–287, 1949.

[19] A.R. Paterson. *A First Course in Fluid Dynamics*. Cambridge University Press, 1983.

[20] L. Prandtl. *Verhandlungen des dritten internationalen Mathematiker-Kongresses (Heidelberg 1904)*. Leipzig, 1905.

[21] L. Prandtl. *Essentials of Fluid Dynamics: With Applications to Hydraulics, Aeronautics, Meteorology, and Other Subjects*. Hafner Pub. Co., 1952.

[22] Lewis F. Richardson. Atmospheric diffusion shown on a distance-neighbour graph. *Proceedings of the Royal Society of London. Series A*, 110(756):709–737, 1926.

[23] P.G. Saffman. *Vortex Dynamics*. Cambridge Monographs on Mechanics. Cambridge University Press, 1992.

[24] Wayne Schubert, Eberhard Ruprecht, Rolf Hertenstein, Rosana Nieto Ferreira, Richard Taft, Christopher Rozoff, Paul Ciesielski, and Hung-Chi Kuo. English translations of twenty-one of Ertel's papers on geophysical fluid dynamics (see the first four papers). *Meteorologische Zeitschrift*, 13(6):527–576, 2004-12-01T00:00:00.

[25] Geoffrey Taylor. The formation of a blast wave by a very intense explosion. i. theoretical discussion. *Proceedings of the Royal Society of London. Series A, Mathematical and Physical Sciences*, 201:159–174, 1950.

[26] Geoffrey Taylor. The formation of a blast wave by a very intense explosion. ii. the atomic explosion of 1945. *Proceedings of the Royal Society of London. Series A, Mathematical and Physical Sciences*, 201:175–186, 1950.

[27] A.A. Townsend. *The Structure of Turbulent Shear Flow*. Cambridge Monographs on Mechanics. Cambridge University Press, 1980.

[28] D.J. Tritton. *Physical Fluid Dynamics*. Oxford Science Publications. Clarendon Press, 1988.

[29] Francisco Vera, Rodrigo Rivera, and César Núñez. Backward reaction force on a fire hose, myth or reality? *Fire Technology*, pages 1–5, 2014.

[30] Th. Von Kármán. Mechanische Ähnlichkeit und turbulenz. *Nachrichten von der Gesellschaft der Wissenschaften zu Gottingen*, 5:58–76, 1930.

[31] G.B. Whitham. *Linear and Nonlinear Waves*. Pure and Applied Mathematics: A Wiley Series of Texts, Monographs and Tracts. Wiley, 2011.

Index